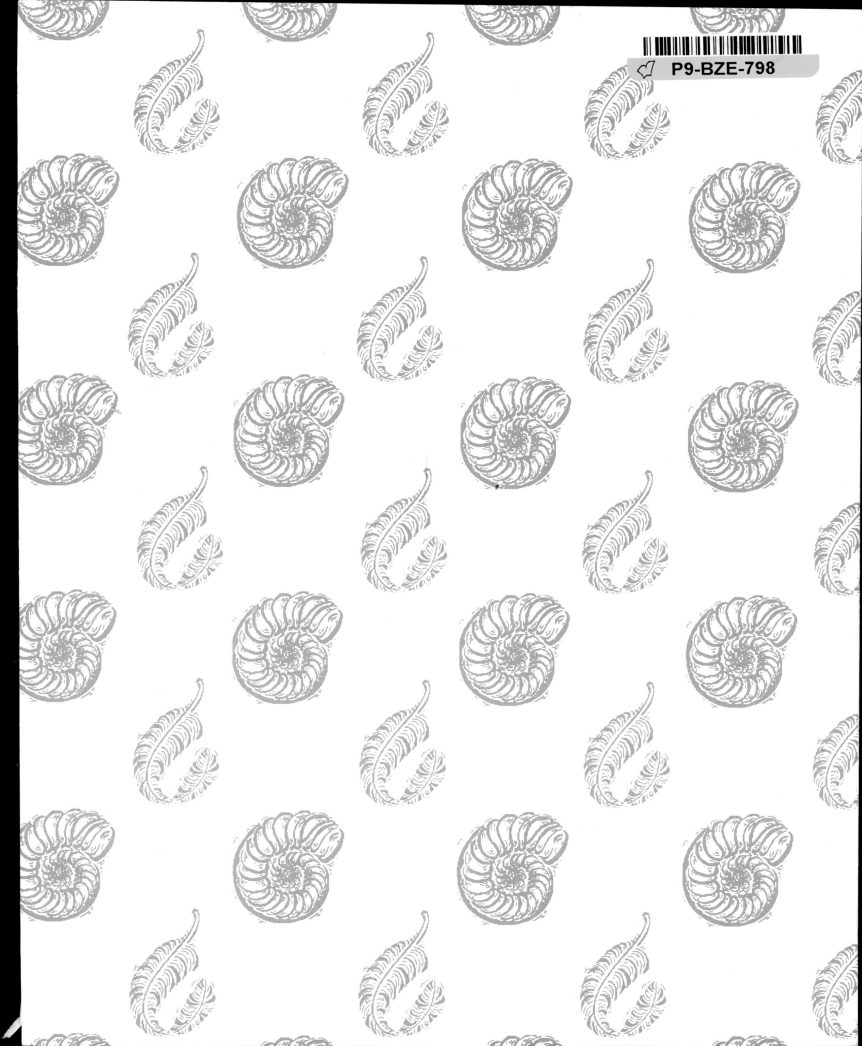

FROM
SO
SIMPLE
A
BEGINNING

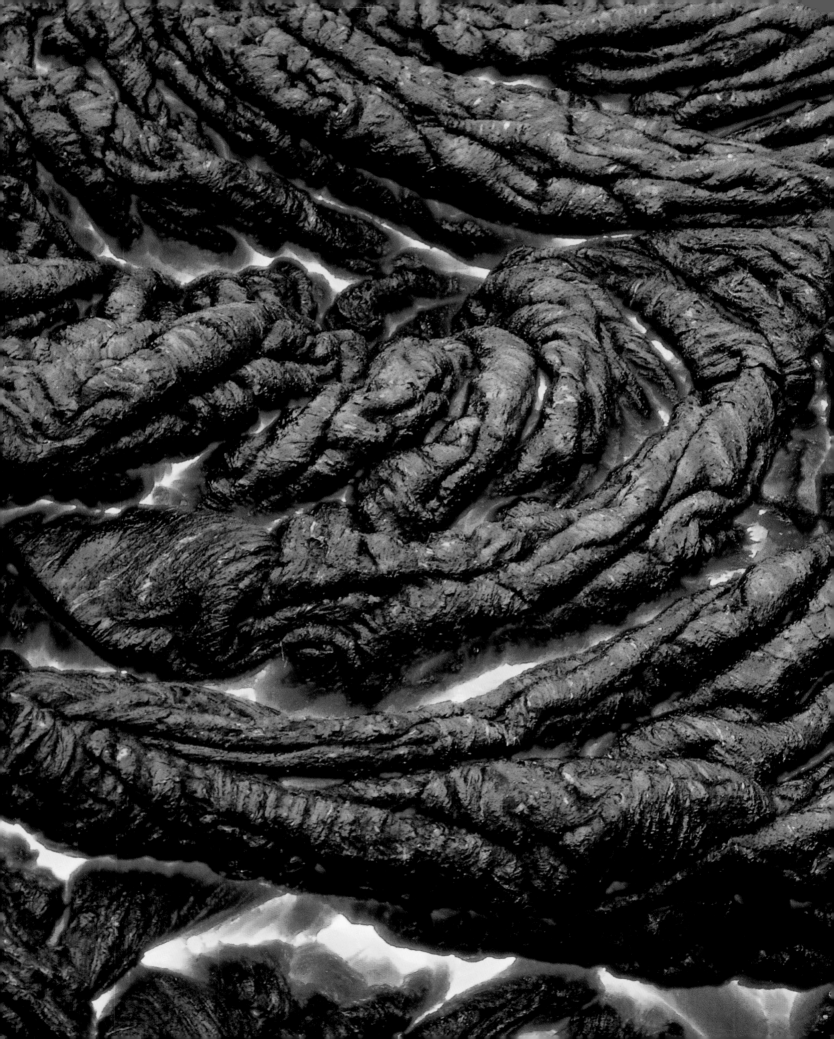

FROM SO SIMPLE A BEGINNING

The Book of Evolution

Philip Whitfield

Macmillan Publishing Company
New York

Maxwell Macmillan International
New York Oxford Singapore Sydney

A Marshall Edition
This book was conceived, edited and designed by
Marshall Editions, 170 Piccadilly, London W1V 9DD

Macmillan Publishing Company
866 Third Avenue
New York, NY 10022

Typesetting by Servis Filmsetting Limited, Manchester, U.K.
Origination by Reprocolor Llovet SA, Barcelona, Spain
Printed and bound in Spain by Printer Industria Grafica, Barcelona

Macmillan Publishing Company is part of the Maxwell
Communication Group of Companies.

Library of Congress Cataloging-in-Publication Data

Whitfield, Philip.
    From so simple a beginning: the encyclopedia of evolution /
  Philip Whitfield.
       p.   cm.
    Includes bibliographical references (p.   ) and index.
    ISBN 0-02-627115-X
       1. Evolution (Biology) 2. Adaptation (Biology) 3. Life
    (Biology)
    I.   Title.
    QH366.2.W53   1993 /   93-6883   CIP
    575—dc20

10 9 8 7 6 5 4 3 2 1

Contents

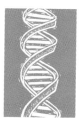

FOREWORD

Adaptation is the most striking fact about life: all organisms seem to be superbly suited to the demands of their daily round, in their physiology, anatomy and behaviour. No matter how small or large, how commonplace or bizarre, each organism has apparently been endowed by nature with the equipment necessary for success in life – or at least for a fighting chance. Ever since the Greeks more than two millennia ago began a serious study of the nature of the world around them, this fact has demanded explanation. To scholars working prior to the establishment of Darwinian evolutionary theory, the answer was obvious: adaptation was an expression of God's work, the result of divine design. The study of God's handiwork in nature was known as Natural Theology. To modern biologists the answer is equally obvious: natural selection, which fits organisms to their environments through the gradual accumulation of small changes in response to the prevailing conditions of the environment.

This explanatory shift – from divine design to natural selection – was the second of two radical intellectual revolutions of the past half millennium. The first, the Copernican revolution of the mid-sixteenth century, displaced the Earth from being the focus of God's attention at the centre of the universe to being one of several planets circling the Sun. The revolution initiated by Charles Darwin (1809–1882), three centuries later, dispelled the notion that the creatures of the Earth – including humans – were not God's immutable creations but were the products of a process of descent with modification, or evolution as it came to be known.

Scholars in the pre-Darwin era had recognized similarities among species, a fact that had enabled the eighteenth-century Swedish biologist Carolus Linnaeus (1707–1778) to construct a detailed and elaborate classification of the major forms of life. (In its essentials, Linnaean classification remains valid today.) Biologists were able to arrange species as a hierarchy, from the "lowest" to the "highest" forms, with *Homo sapiens* occupying the topmost position, a little lower than the angels. This Great Chain of Being represented a static expression of God's creativity. For scholars in the post-Darwin era, similarities among species are an expression of a shared evolutionary relationship, with all species ultimately deriving from a single common ancestor (or a few). The concept of descent with modification therefore transformed the static Great Chain of Being into an historical record of a dynamic process of evolution.

When Darwin published his *Origin of Species* in November 1859 he explained his hopes as follows: "I had two distinct objects in view; firstly to show that species had not been separately created, and secondly, that natural selection had been the chief agent of change." He succeeded immediately with the first, while the second took longer. Natural selection did not come to be widely regarded as the chief agent of evolutionary change until the early 1940s, long after Darwin's death.

Darwin was by no means the first to publish on the subject of evolution, usually termed the transmutation of species. His grandfather, Erasmus Darwin (1731–1802), wrote extensively on the notion of relatedness through inheritance among species and speculated on mechanisms of transmutation. His ideas were taken up by the French biologist Jean Baptiste de Lamarck (1744–1829), whose theory of the inheritance of acquired characters was very popular, and intrigued the young Darwin. Many books and articles on other possible mechanisms of transmutation were penned in Europe and the United States during the first half of the nineteenth century, some themes of which came very close to that of natural selection. And Darwin was very nearly preempted when, in 1858, the British naturalist Alfred Russel Wallace (1823–1913) also hit on natural selection as the agent of transmutation.

The notion of evolution was therefore very much in the air when *Origin* was published, and its thesis found a receptive audience in the scientific community, if less so in religious circles. The book, which Darwin described as "one long argument", was a comprehensive compilation of facts, from observations of natural history, geology, embryology, paleontology and the domestication of plants and animals. The weight of the evidence was unassailable: transmutation was accepted as a fact. Natural selection, which is based on the inheritance of favourable genetic variation, was, however, viewed with scepticism. One reason was that very little was known at the time about the mechanisms of genetic change and inheritance.

In addition, natural selection was perceived – correctly – as producing a purposeless path of evolution, with biological change blindly tracking environmental change. During the late nineteenth and early twentieth centuries, most biologists strongly favoured progressive modification, with evolution leading steadily to higher, more complex, more advanced forms of life. In this intellectual atmosphere, blind chance as a key element in the engine of evolution was philosophically unacceptable. Directed, progressive evolution was known as orthogenesis.

The Austrian monk Gregor Mendel (1822–1884) laid the foundation of modern genetics with his breeding experiments on peas, in 1865. His work showed that the inheritance of characters – such as colour and shape – was atomistic, that is, determined by discrete genetic entities. Mendel's insight was overlooked for four decades, however, to be rediscovered early this century, when it was immediately used to promote a rival theory to natural selection. Evolution proceeded by macromutations, or large jumps, argued the prominent geneticists Hugo de Vries (1848–1935) and William Bateson (1861–1926), not by incremental changes as required by natural selection.

During the 1930s, however, the mathematical

treatment of Mendelian genetics by the British scholars Ronald A. Fisher (1890–1962) and J.B.S. Haldane (1892–1964) and the American Sewell Wright (1889–1988) showed that the inheritance of discrete genetic units, now known as genes, was compatible with the continuous variation of characters seen in populations. Darwin's theory now had what it had lacked for half a century: a foundation in a well-tested theory of inheritance. This mathematical insight, in combination with a wider understanding of population biology, rescued Darwin's key agent of evolutionary change, and natural selection became the centrepiece of modern evolutionary theory. The publication in 1942 of a book by Julian Huxley (1887–1975), titled *Evolution: the Modern Synthesis*, served as a landmark for the beginning of the modern theory, also known as neo-Darwinism.

So powerful was neo-Darwinism that it became a theme for uniting all of biology. The incremental changes that are the stuff of natural selection came to be viewed as the source of all evolutionary change, from slight modifications in, say, a species' colour to the origin of major novelties, such as the emergence of the mammalian reproductive system from a reptilian precursor. In this view, large changes were the same as small changes, extrapolated to a grander scale.

Perhaps it was inevitable that so sweeping a view should be challenged. So it was that in 1972 the American paleontologists Niles Eldredge and Stephen Jay Gould argued that natural selection – as expressed in neo-Darwinism – was inadequate to explain the evolutionary pattern seen in the fossil record. Species do not change gradually and continuously throughout their existence, but tend to remain unchanged once they have evolved, and then disappear or change rapidly after a long period of time. The pattern of change, known as punctuated equilibrium, implied that, central though natural selection may be to evolutionary change, its effects are constrained in important ways. For instance, pathways of embryological development may be limited, thus reducing the range of possible body forms. Change, when it is possible, is therefore rapid.

Considerable debate ensued over the reality of the pattern described by Eldredge and Gould, and about underlying mechanisms. The fossil record is notoriously incomplete, and may simply give the impression of a punctuated pattern, some argued. During the past decade this has been examined in detail, revealing that evolutionary change is sometimes gradual and sometimes punctuated. Part of the debate is whether new species are more likely to arise as a result of gradual or punctuational change, but this remains unresolved.

As a result of this debate and the research it stimulated, neo-Darwinism has been extended and enriched. Natural selection remains at its core, but with other factors (including developmental constraints) influencing possible outcomes, both at the level of individuals within a species and of differential success among species.

In Darwin's time, fossil evidence was limited and the record extremely incomplete. For various theoretical reasons, Darwin baulked at the notion of occasional mass extinctions, and argued that a more complete record would show them to be artefacts, illusions created by gaps in the record. In recent years, mass extinctions have been shown to be a real part of the record, and they have a striking effect on the history of life.

The mechanism of natural selection implies that a species' success is determined by how well it is fitted to prevailing circumstances, including its interaction with other species – the struggle for existence, as Darwin put it. A species that fails to compete may become extinct. When mass extinctions occur, however, these rules change. Whatever their cause – whether through global climate change or asteroid impact – mass extinctions elect as their victims species with characteristics having nothing to do with everyday success or failure. For instance, large-bodied species are more vulnerable to mass extinctions, as are species with limited geographical ranges.

When mass extinctions occur, therefore, many species will disappear, regardless of the suitability of their adaptation. The evolutionary clock is therefore reset, with new species arising from the lucky survivors. This new insight into the workings of mass extinctions adds a further element of chance into the overall shape of the history of life.

Darwin's insight into what was necessary to demonstrate the fact of evolution and the principal mechanism by which it came about was remarkable, and remains the unifying theme of modern biology, from molecular biology and genetics to the most complex of ecosystems. This book captures the power of that unifying theme, and invokes the final paragraph of the *Origin*:

> *"There is a grandeur in this view of life, with its several powers, having been originally breathed by the Creator into a few forms or into one; and that, whilst this planet has gone cycling on according to the fixed law of gravity, from so simple a beginning endless forms most beautiful and most wonderful have been, and are being evolved."*

Like Darwin, we stand in awe at the wonderful creativity of nature, with an understanding of its laws enhancing its beauty for us, not diminishing it.

Roger Lewin
Washington, D.C.

LIFE AND CHANGE

Evolution is something that happens to living things. It is the alteration of life through time. One sort of evolutionary change is defined as adaptation – the slow, generation by generation alteration of the mix of characteristics which typify a species. The members of a certain snail species, for example, may have more stripes on their shells than their predecessors of a few generations before. Members of a mouse species might have longer hair than their immediate ancestors.

Slow, inherited changes of this sort may well be useful. Snail shells with more stripes may improve camouflage and thus reduce predation by thrushes. As the climate cools over the course of decades, longer hair may decrease heat loss from a mouse's body and improve its survival chances.

Such alterations are known as adaptations because they appear to make the organisms concerned better suited, better adapted, for the particular conditions in which they live. The changes of adaptation take place within the populations of a single species. Potentially, at least, all the members of that species can breed with one another, despite minor differences between individuals.

The record of the living past provided by fossils tells a story of new forms of animals and plants arising, flourishing and after a time becoming extinct. It points to the existence of more radical evolutionary change leading to the creation of new species – a process known as speciation. If the range of inherited adaptations in one population of a species becomes so extreme that it precludes breeding with other populations, the changed population is effectively a new species, a whole new type of organism.

Speciation and adaptation are facts of life, but the study of evolution tries to explain why they happen as they do. At the end of the 18th and beginning of the 19th centuries, there was a growing realization that living things did change. Ideas about what drove those changes crystallized into a formal theory through the insights of Charles Darwin and Alfred Russel Wallace who, in the mid-19th century, both suggested that the dynamism of life and its evolutionary changes were the result of natural selection.

The theory of natural selection is probably best explained as a series of statements and effects. First, each generation of a species produces more offspring than will survive to breed in the next generation. Second, individual members of a species in a generation can all potentially interbreed, but they will not be identical. Third, some variable characteristics will give the individuals that possess them a competitive edge in the struggle for survival and breeding chances, which will in turn enable them to produce the next generation.

Assuming that those advantageous characteristics are inheritable, they are more likely than less advantageous ones to show up in subsequent generations. They have been naturally selected. As a result, the species will change through time in ways that fit its members better for the environment in which they struggle to survive.

This sequence of causes and effects produces a pattern of change recognizable as adaptation. It can be compressed into epigrams such as "the survival of the fittest", but such abbreviations miss much that is important – for example, the intrinsic and inevitable nature of variation. These generalizations obscure the fact that the mix of variation in every generation of every living plant and animal is tested against the conditions in which those organisms live. They also blur the crucial condition that advantageous characteristics must be inheritable and inherited for natural selection to occur.

Although 150 years old, this Darwinian theory of evolution is arguably still the most profound text that can be written about the nature of life. Its essence still underpins all modern life sciences. New knowledge about the molecular basis of genetics and life simply enables us to redefine Darwin's ideas in ever more precise language.

THE EVOLUTION OF EVOLUTION

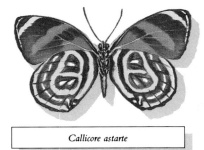

Callicore astarte

Callicore sorona

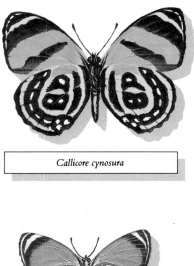

Callicore cynosura

Diaethria aurelia

Much impetus for evolutionary thought in the 19th century came from the sudden availability of new biological knowledge from around the world. The collecting of butterflies, beetles and other creatures became popular and vast numbers of new species were described.

Charles Darwin's book on evolution, *On the origin of species by means of natural selection or the preservation of favoured races in the struggle for life*, was an instant sensation. The first edition of 1,250 copies sold out on the first day of publication in 1859.

That year can justifiably be said to mark the start of the age of modern life sciences. Darwin's insights brought together knowledge from a dozen fields and produced a single framework for the understanding of life. From 1859 onward, all serious speculations about life on Earth took place in the context of Darwin's theories on evolution.

This portrait of Darwin was painted by George Richmond in 1840, four years after Darwin's return from his voyage on the *Beagle*.

Given the status and influence of the theory of evolution today, it is hard to imagine its anguished birth about 150 years ago in Victorian England. The evolution of the theory of evolution is a story that combines outstanding scientific discovery, powerful personalities and the constraints once imposed on science by society and religion.

Both before and during the time of Charles Darwin, the orthodox Christian view measured the history of Earth in only thousands of years. It demanded literal acceptance of the origin of all living things as described in the Bible in the book of Genesis – they were the special and one-off product of the creative powers of the deity; human beings were the apogee of that creative process.

Any suggestion that living things changed through time could not be countenanced: it would imply, blasphemously, that the life forms created by God were in some way imperfect. The idea that life did not change also implied that all the initially created life forms were still on Earth. Why would God create an animal or plant only to let it become extinct later?

These orthodoxies were under assault long before Darwin's exploratory trip on HMS *Beagle* during the 1830s. Freethinkers of various persuasions had questioned the traditional position. Some had philosophical doubts, others were influenced by new types of evidence that accumulated in the 18th and early 19th centuries – particularly evidence relating to the nature and significance of fossils.

In many parts of the world, fossils were so numerous that they were impossible for even non-specialists to miss: they studded cliff faces and lay in profusion on scree slopes and beaches. Shells, teeth, coiled ammonites and bones, all made out of rock, were found during this period. Early ideas on their origin were vague and diverse. Were these objects the remains of creatures or something quite different? Some believed that the shaped stones were the workings of a life force in the Earth, straining to make images of the creatures of God's creation. Others suggested that the eggs of real animals had lodged in the rocks and developed as rocky tumours.

But the links between fossils and existing animals became more and more obvious. Eventually, there could be no doubt that the fossils were the petrified remains of previously living things. In order to explain this fact in religious terms, the Flood described in the Bible was invoked. Fossils were said to be the remains of creatures drowned in the deluge, locked for ever in sediments produced during the inundation. The theory also explained why fossils of sea creatures were found in the rocks at the tops of mountains: a gigantic flood could have swept shells and other remains from the sea to these heights.

New problematic facts began to emerge, however. Geological studies revealed that different layers of sedimentary rock could not all have been formed at once. And the time required for their formation seemed to be much greater than a few thousand years. If all the fossils and layers were not the result of a single biblical flood, they could have been the product of a sequence of floods, of which Noah's was the last. But these earlier catastrophes were not recorded in the Old Testament. Even more worrying for traditionalists was the realization that some fossils were of creatures that no longer lived on Earth. Rather they seemed to represent life forms that had long been extinct.

For a while this difficulty was answered by the suggestion that the creatures represented in the fossils still flourished in unexplored parts of the globe. As ever more remains of extinct types accumulated and the pace of global exploration quickened, however, this defensive position was eroded. If God had created all his creatures at one instant in the past, he had subsequently let some of them die out.

Change in living things was thus a matter of heated debate and controversy at the end of the 18th and beginning of the 19th centuries. One concrete expression of this debate was the work produced by a member of the poorer end of the French nobility, Jean-Baptiste Antoine de Monet (1744–1829), who styled himself Chevalier de Lamarck and has been known ever since simply as Lamarck. His thoughts on the organization of living things and their capacity for change are dismissed today, but unjustly so. In fact, Lamarck produced in 1809 what has been described as the first explicit theory of evolution.

Lamarck was a skilled field naturalist and a masterly botanist. He invented the immensely successful dichotomous key system for identifying an unknown animal or plant from its observable characteristics. The system is a set of questions, each bracketed with two or more possible answers which send the enquirer to yet another question of the same type. At the end of each chain of questions and answers is the identification of a creature.

A hypothetical example of Lamarck's system could identify a kiwi by answers to two questions. First, is the bird able to fly or flightless? If flightless, is its beak more than twice the length of the rest of its head? The only flightless bird with a beak that long is the kiwi. Lamarck's simple system has stood the test of time and remains the standard method for codifying identifications.

Lamarck became the Royal Botanist at the Jardin du Roi in Paris, but remained a challenger of the orthodoxy in his analysis of living things. His taxonomic surveys showed him that groups of species were related to one another in meaningful ways and that species did change through time. His ideas on these and related themes were first laid out in his book *Philosophie zoologique*, published in 1809.

Lamarck thought that different types of animals, with their varying levels of complexity, were part of a long-term

Fossils, such as this 50 million-year-old fish, stimulated evolutionary thought in the 19th century. Once it became known that many fossils were of long-extinct creatures it was obvious that life could, and did, change through time.

Callicore maimuna

Callisthenia markii

Marpesia marcella

Catonephele numilea (male)

series of transformations. The simplest creatures were assumed to arise by "spontaneous generation" from non-living things – at that time an uncontroversial assumption. They then moved up through the levels of increasing complexity toward the perceived perfection of humans. Nearly 200 years ago, in the face of religious certainties, this was revolutionary thinking indeed.

Much of the rest of Lamarck's "zoological philosophy" was weakened by lack of evidence on the mechanisms of these transformations. Lamarck suggested that the animals moved up the chain because they wanted to be better; their desire for improvement was supposed to fuel the change. For example, giraffe ancestors had short necks, but by stretching up for higher, more juicy leaves, their necks lengthened. According to Lamarck, their offspring were then born with longer necks as a consequence of the desires of their parents.

Lamarck's linking of desire for change with change itself was soon ridiculed and

he has been vilified ever since. Common sense alone shows that changes acquired during a lifetime are not passed on to young. Jews have been circumcising their sons for thousands of years but they are still born with foreskins.

Although flawed, Lamarck's theory about the giraffes did recognize important facts: that the ancestors of the giraffes had short necks which subsequently became longer. His assertions removed God as the instigator of all life forms at one moment in time and showed that not only do animals change through time, but also that these changes make them better adapted to their surroundings. Lamarck recognized these processes but failed to identify their mechanisms.

It was left to two other key figures, with differing reputations, to explain this engine of change. The name of Charles Darwin is firmly welded to the idea of evolution, while Alfred Russel Wallace is often seen as a bystander. In fact, these two men independently gave birth to the same explanation for

Undisturbed layers of sedimentary rock, such as these Cretaceous examples from Norfolk in eastern England, enable the fossil-hunter to trace the past. In general, the lower rock layers are older than those above, and the fossils in them will have the same age pattern.

Lamarck, a controversial 19th-century naturalist, believed that because the giraffe stretched up to reach leaves it evolved a long neck. But although the okapi, a relative of the giraffe, stretches in the same way, it has persisted with the shorter neck seen in fossil giraffes. This demonstrates that evolution is not driven by simple patterns of use and non-use.

the motive force of evolution – the theory of natural selection.

Charles Darwin was born on 12 February 1809, the son of a successful physician. His grandfather, Erasmus Darwin, also a physician, was a Lamarck sympathizer and wrote a long poem entitled *Zoonomia*, which contained shadows of evolutionary ideas. Charles lived among the moneyed English upper and middle clases. In 1839 he married his first cousin, a member of the Wedgwood family which ranked high in the country's industrial elite.

Originally destined for life in the church, Darwin studied at the University of Edinburgh and then at Cambridge. From his early youth he was a fanatical naturalist with a particular fascination for beetles. His leaning toward the collection and identification of living things prompted the crucial decision that was to change not only his life but also the course of life sciences. The decision turned an apparently ordinary Englishman into an extraordinary scientist, one

who, arguably, has transformed our view of ourselves more than any other.

Despite strong family pressures to continue his progress toward life as a country parson, Darwin agreed to accompany Captain Fitzroy, commander of the 10-gun brig HMS *Beagle*, on a five-year survey voyage around the globe. His sponsor was one of his Cambridge professors, the botanist J. S. Henslow. Henslow explained that Captain Fitzroy was after a socially

congenial companion for the voyage as well as a naturalist. Darwin came from the right social background and was an avid collector.

The *Beagle* sailed from England in 1831, two days after Christmas. Its route took it round the east and west coasts of South America and to the Galápagos and other islands in the Pacific before returning in 1836. The findings and experiences of this voyage – the only serious travels Darwin ever made – had profound

Catonephele acontius (male)

Catonephele acontius (female)

Myscelia cyaniris

Myscelia orsis

effects on his thinking. They convinced him that living things had patterns of linked ancestry.

In the journals and notebooks of the voyage itself and in the records of his studies after the voyage, Darwin grasped with increasing clarity that the shared features of similar creatures spoke more of common origins than of the Divine Plan of a Creator. He saw, for example, the bone structure of most mammalian limbs – one long bone, then two long bones and a five-digited hand or foot – as evidence that all mammals were derived from a single ancestor with those particular characteristics.

Darwin took the giant sloth fossils he found in South America as evidence of extinctions, since their only living relatives were much smaller species. He noted that animals could possess apparently functionless "vestigial" organs. He argued that the multiple vertebrae in a bird's tail, whose role for feather attachment could equally well be served by one short vertebra, could only be sensibly explained if birds had arisen from reptile ancestors that had long tails with many vertebrae. Why should a creator put many bones into a bird's tail when one would have functioned as well? Then, on the archipelago of the Galápagos Islands Darwin found many apparently closely related finch species. He believed that they were all descendants of a single ancestral type that had managed to colonize the islands from the South American mainland far to the east.

During the voyage Darwin had hundreds of specimens shipped back which were later analysed by himself and other naturalists. He became convinced that evolutionary change in all living things was a fact – a realization that took him far from the religious constraints of the country parson he was to have been. Darwin became a biological rationalist.

Darwin also confronted the issue of a mechanism for change in species. In this he was influenced by the demographic theories of the Reverend Thomas Malthus (1736–1834). In his *Essay on the Principle of Population*, Malthus had argued that since human populations can increase faster than supplies of food, starvation and competition for food will ultimately result, producing a regulatory controlling force on population growth.

In his autobiography Darwin noted "In October 1838 . . . I happened to read for amusement Malthus on population . . .". It was a fancy that led far. Darwin took a germ of an idea from Malthus and built it into a general theory about the dynamics of life.

The talented naturalist Alfred Russel Wallace (1823–1913) made his living by travelling the world, collecting specimens of birds and butterflies for museums and private collections. He became increasingly interested in the nature of evolution and, after extensive travels in the Amazon and Malaysia, wrote a paper titled "On the law which has regulated the introduction of new species".

Wallace knew he still had not found the key to the mystery of evolution. But three years later while working in Southeast Asia he had an inspiration, triggered by his reading of Malthus's work on population, that evolution was driven by natural selection. He did not know that, quite independently, he had reached the same conculsion as Charles Darwin working on the other side of the world in England.

At its simplest the theory was as follows. As the members of a species breed there is an inevitable increase in numbers and thus competition for crucial resources such as food. Not all individuals are equally well equipped for the struggle – there is always variety within any species. Those that are slightly better adapted to gain resources in a particular environment tend to survive and breed at the expense of those less well equipped.

If the useful characteristics of the survivors are passed on to their offspring, the species will gradually change to become better adapted to its surroundings. This is the concept of natural selection: environmental stresses, such as competition for resources, cause individuals to be "selected" from among the variability within a species. Selection in

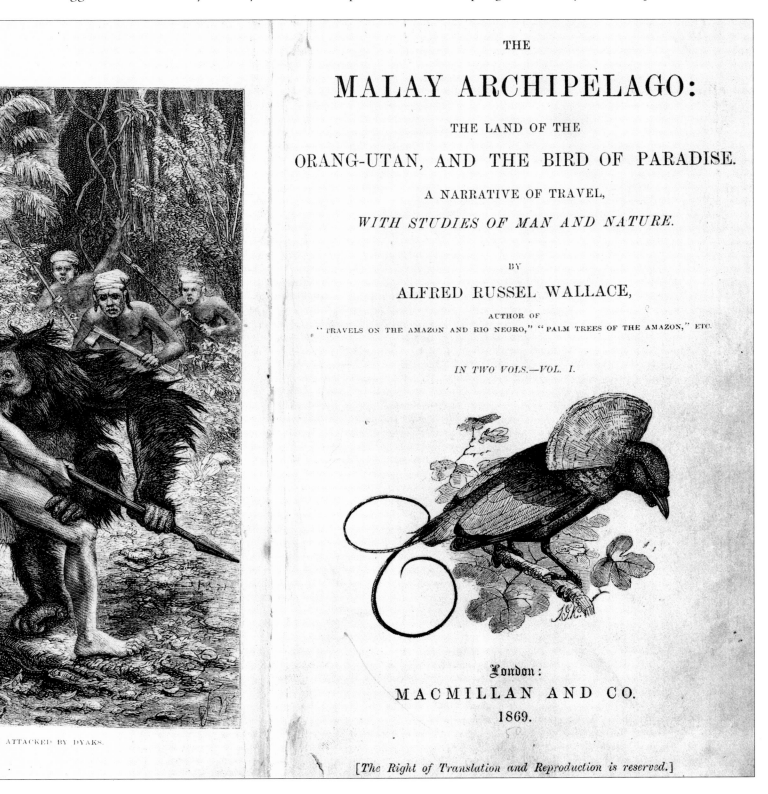

TAN ATTACKED BY DYAKS.

THE

MALAY ARCHIPELAGO:

THE LAND OF THE

ORANG-UTAN, AND THE BIRD OF PARADISE.

A NARRATIVE OF TRAVEL,

WITH STUDIES OF MAN AND NATURE.

BY

ALFRED RUSSEL WALLACE,

AUTHOR OF

"TRAVELS ON THE AMAZON AND RIO NEGRO," "PALM TREES OF THE AMAZON," ETC.

IN TWO VOLS.—VOL. I.

London:

MACMILLAN AND CO.

1869.

Eunica sophonisba

Biblis hyperia

Haetera macleannania

Cithaerias esmeralda

this sense simply means that the inherited characteristics of certain individuals make them more likely to survive and breed in that set of environmental conditions than other individuals.

Darwin's model for natural selection was the artificial selection used by animal and plant breeders. Their selection of particular characteristics from generation to generation can produce a bulldog or a whippet from a wolflike ancestor, or a plump-eared wheat from scrawny wild grass. By studying highly bred or "fancy" animals such as pigeons, Darwin was able to draw a direct analogy between such artificial selection and that achieved by competition in the natural environment.

The skeleton of all these ideas was in Darwin's mind by 1838, but he did not publish them until 20 years later. One reason for this delay was probably a desire to buttress every hypothesis with facts to avoid the fate of Lamarck's earlier theories. Another reason for delay was almost certainly the shock waves he knew such a theory would cause in his society and its religion. In his own notes Darwin had confronted the ultimate ramification of a theory of evolutionary change and natural selection; he knew that it evoked a God-less landscape for life. It also placed the human species squarely within that biological landscape as merely the temporary endpoint of a line of ape evolution.

The block, induced by his desire for scientific rigour and perhaps even greater anxiety about the public consequences of disclosure, was finally sundered by the fear of his work being "scooped" by Alfred Wallace.

Wallace, born in 1823, was 14 years younger than Darwin. A highly successful naturalist, he was a professional collector, not a moneyed amateur like Darwin. He made extremely productive collecting trips to Amazonia and the East Indies and sold parts of his collections to support himself. While in the East Indies he had a blinding flash of inspiration about the nature of life. In terms that differed little from those in Darwin's then unpublished notebooks, Wallace conceived the idea of natural selection. As he put it, "species will tend to depart indefinitely from their original type".

Wallace was an admirer of Darwin and had been inspired by his account of the *Beagle* voyage. So he sent his hero Darwin details of his own theories on how species might change through time. Darwin was dumbfounded. Wallace's work seemed likely to preempt all his own years of painstaking data gathering and analysis. An extraordinary compromise, brokered by Darwin's fellow scientists including Joseph Hooker and Charles Lyell, was then settled upon. It was arranged that papers by both Wallace and Darwin on the topic at issue should be given at the same meeting of the Linnaean Society in London.

The following year, 1859, Darwin published *On the Origin of Species by Means of Natural Selection*, his detailed volume on the evidence for evolutionary change and natural selection. The reception that the book received led to Darwin being regarded as the parent of evolution and Wallace as a bystander at the delivery. The reality was different.

Within a few years of the publication of Darwin's book the basic fact of the evolutionary change of living things had been firmly established, despite the continuing attack of those who clung to the less frightening, pre-Darwinian world view. Acceptance of the idea of natural selection driven by the inheritance of favoured characteristics came later. It was not until the mid-20th century that the genetic basis of heredity, based on the plant breeding experiments of Gregor Mendel (1822–84), flowered into a science that could explain variation and its inheritance in the context of an evolving population of individuals.

By the 1950s, a neo-Darwinian consensus tied genetics firmly to evolutionary change. Subsequent developments in molecular biology have greatly expanded the consensus. Once the sheer molecular scale of genetic information becomes clear – namely that the genes in a single cell can contain billions of separate instructions – the near inevitability of inheritable changes follows.

Genes have to be copied every time a plant or animal reproduces, a process which takes place in many steps, and it is impossible for this to happen always without error. These errors in the copying of genes are in fact the raw material of evolutionary change. Similarly, the way in which organisms adapt to their environments through evolution can also be seen to be inevitable.

Genetic changes, which can be thought of as fortuitous errors, only pass into the next generation if they are in organisms that succeed in reproducing themselves. If the altered genes do anything to enhance the chances or extent of reproduction, they increase the odds that they will occur in the next generation.

Any genes that diminish reproductive success have the reverse effect. The result is a simple linkage between the fate of a genetic change in a creature and its effect on reproduction. The link keeps changes broadly beneficial with respect to an external world that is constantly in flux.

From our vantage point, near the end of the 20th century, the ideas of evolution can now be seen as a fundamental platform of understanding. The concept of evolution is all-pervasive. It inter-relates with every facet of modern life sciences, from molecular biology to anthropology, which makes it difficult to isolate for inspection a single entity called evolution. Evolution is no longer a separate and distinct theory; it is something that colours all our thoughts about living things on this planet.

Analysis of the genetic basis of inherited characteristics helps to explain the underlying mechanisms of evolution. Many of the early examples of this type of analysis were based on plant-breeding experiments in which single genes control obvious characteristics of the plants or their seeds. Mendel's early work, for example, was carried out on peas and much subsequent research has used maize or sweetcorn.

This simple experiment shows the result of a cross between a maize plant grown from a seed on a cob with purple seeds and another grown from a seed on an all-white cob. Because of the genetic constitution of the parent plants, half the seeds of the offspring cobs are purple and half white. One parent had a purple and a white gene but the effect of the purple dominated. The other had two white genes. This starting point produced equal numbers of purple and white seeds.

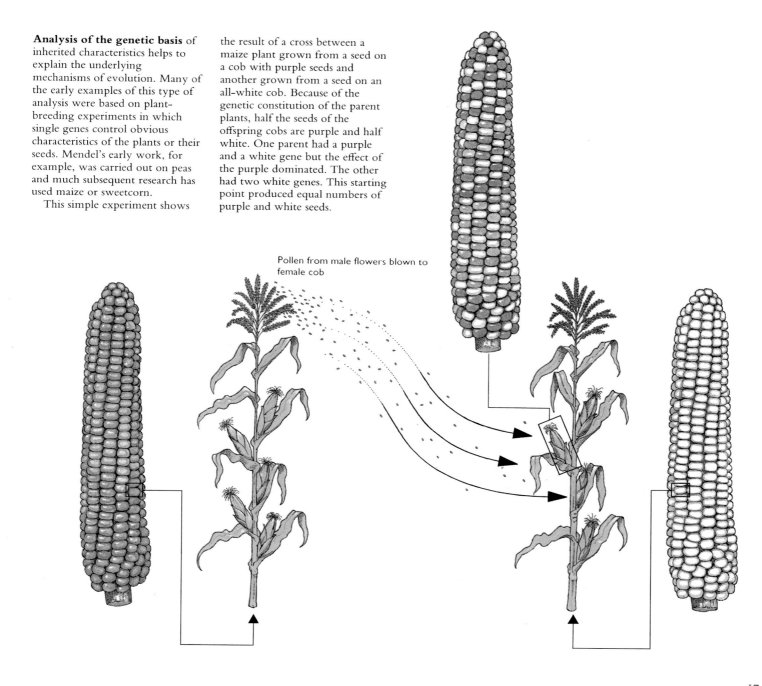

Pollen from male flowers blown to female cob

THE TIME SCALE OF LIFE ON EARTH

Blue-green algae, like modern plants, gave off oxygen during photosynthesis.

The fossils known as stromatolites were built up by mineral deposition in layered mats of blue-green algae and bacteria in the sea.

Stromatolites

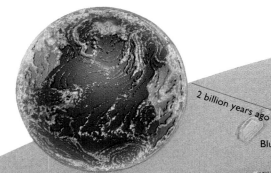

2 billion years ago

Blue-green algae

The Milky Way galaxy

6 billion years ago

7 billion years ago

8 billion years ago

9 billion years ago

Our solar system

Earth's atmosphere was slowly made aerobic by the oxygen produced by photosynthetic blue-green algae.

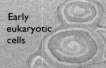

Early eukaryotic cells

5 billion years ago

Simple clusters of cells with nuclei (eukaryotes) were present by 850 million years ago.

The Earth is believed to have been formed by a process of accretion round the sun at the beginning of the solar system.

1 billion years ago

Toward the end of the Precambrian – the long period from the formation of Earth until the beginning of the Cambrian – a number of larger, more complex animals developed. Best known of the fossils from this period are those found in the Ediacara Hills in Australia, but fossils similar to those of the Ediacaran fauna have now been discovered elsewhere in the world. The affinities of these creatures are uncertain. They could represent early versions of modern groups such as sea pens and annelid worms. Alternatively, they could be unique early multicellular animals – metazoans – unrelated to modern groups.

Cambrian 570 million years ago	**Ordovician** 500 million years ago	**Silurian** 440 million years ago
At this time there was a sudden flowering of complex animal forms, such as molluscs and arthropods, many with hard shells or exoskeletons.	The animals of the Ordovician were still all marine. Trilobites, members of the arthropod group, were particularly numerous.	Top predators in the sea at this time were the huge sea scorpions, but the first jawed fish were also developing. The land was colonized by plants and later by invertebrate animals.

Ediacaran animal (*Tribrachidium*)

Burgess Shale invertebrate (*Aysheaia*)

Trilobite (*Encrinurus*)

Sea scorpion (Eurypterid)

Earth was formed about 4.5 billion (4,500 million) years ago. There has long been controversy about this date and therefore the real age of our planet, but in the last few decades a mixture of analytical techniques, most based on measurements of the amounts of different atomic isotopes in the Earth's rocks, have allowed this consensus view. All the planets, moons, asteroids and other space debris of the solar system were formed at almost the same time as Earth – the oldest rocks found in meteorites and on the Moon are also about 4.5 billion years old.

Evolution takes time. When Earth was formed it was a lifeless planet among other lifeless planets circling our sun. It was not until about one billion years later that there were life forms of sufficient complexity to leave their microfossil traces in rocks of the time. But the real origins of life – the origins of self-replicating chemical entities – must have predated those earliest fossils by a considerable period. Earthly life has evidently been moulded by between 3.5 and 4 billion years of evolution.

All the earliest forms of cellular life on Earth were simple cells without nuclei (prokaryotes) that lived in water. These included different forms of bacteria.

3 billion years ago

Orbiting solid fragments – particles and lumps of metals and rocky materials – gathered under the influence of gravity.

Radioactive heat melted the planet's core. The dense metallic materials settled at the centre and lighter rocky ones at the surface.

The surface of the Earth cooled to form a crust – still bombarded by meteors. The first atmosphere formed.

Tertiary 65 million years ago
After the loss of the large reptiles of land, sea and air at the end of the Cretaceous, mammals and birds became the dominant life forms.

Quaternary 2 million years ago
Early ape-men and humans evolved in Africa and spread all over the world.

Cretaceous 140 million years ago
On land, flowering plants evolved, as did the bees and butterflies that pollinate them. Many more plant-eating and carnivorous dinosaurs developed.

Triassic 245 million years ago
Reptiles dominated the world. Turtles and ichthyosaurs thrived in the seas, pterosaurs triumphed in the skies, while the first dinosaurs – and mammals – walked the land.

Permian 290 million years ago
Plant life on land was dominated by conifers. Reptiles began to replace amphibians as the dominant land animals. Trilobites became extinct.

Devonian 410 million years ago
In the seas a remarkable diversification of jawed and jawless fish took place. Amphibians came on to land and early insects developed.

Homir

Early elephant (Gomphotherium)

Carnivorous dinosaur (Tyrannosaurus)

First known bird (Archaeopteryx)

Land horsetail plant (Calamites)

Jurassic 210 million years ago
Cycads, ferns and tree ferns clothed the land in this period. Many types of herbivorous and carnivorous dinosaurs evolved – and the first true bird, Archaeopteryx. The seas teemed with bony and cartilaginous fish as well as marine reptiles and coiled ammonite molluscs.

Carboniferous 365 million years ago
Swampy forests of large horsetails, club mosses and gymnosperms covered much of the land. The first small reptiles evolved alongside flying insects such as cockroaches and dragonflies.

Early amphibian (Seymouria)

Primitive bony fish (Thursius)

Early reptile (Askeptosaurus)

FOSSILS: THE FABRIC OF EVIDENCE

These mysterious "tramlines" in Jurassic rock from Dorset, England, are the fossilized trails left by an unknown crustacean animal 200 million years ago. The living animal would have wandered over soft, silty mud, leaving two parallel furrows in its wake. These marks have been transformed into a direct trace fossil.

Entombed for ever in fossilized amber are these two 30-million-year-old flies. The amber itself was formed from resin that exuded from trees. Sticky drops of resin on a tree trunk acted as living "fly paper", trapping small insects and spiders. The amber eventually hardened to rock and the creatures remain perfectly preserved in their transparent tomb.

Fossils of bones, shells, even whole animals, make time travel a reality. Looking at them, people today can "see" into the past. Once it was accepted that these "shaped stones" really were what they seemed to be — the petrified remnants of prehistoric creatures — their significance for the story of evolution became clear. Paleontology — the study of fossils — provides answers to many key questions about evolution.

The evidence of fossils proves beyond doubt that the animals that swam in ancient seas and the plants that composed prehistoric forests were different from those of today. Fossils of organisms no longer living on Earth demonstrate that

creatures have become extinct. Chronologically dated sequences of fossils from successive layers of rock reveal patterns of change in groups of living things.

A fossil bone — a bone that has been turned to stone — is one of the most familiar types of fossil. Produced by direct skeletal fossilization, it is a positive mineralized replica of a part of a living skeleton. The reassembled fossil bones of a giant dinosaur in a museum help people now to gain an impression of the size and appearance of an animal that splashed through swamps 100 million years ago. But fossil bones are just one of an extraordinary diversity of fossil types from which an inventory of the ancient

life of our planet can be constructed.

The harder and more resistant to damage and decay a structure is, the more likely it is to be preserved in fossil form. Thus the teeth of vertebrate animals are common fossils. Some invertebrate creatures have shells and other tough outer coverings instead of bones and these, too, are often fossilized. There are plentiful fossils of the flattened shells of bivalve molluscs, such as mussels, and the plated exoskeletons of different types of echinoderm, such as sea urchins.

Unless distorted by pressure, heating, or other physical forces in the rocks, a direct skeletal fossil retains much of the shape of the living creature (or part of creature) from which it formed. In addition, the matrix rock, the rock in which the direct fossil is embedded, often takes on a negative imprint from the positive fossil. These negative impressions can often be just as useful as the positive fossils, providing details about the surface appearance of an animal or plant.

Fossils can take on other forms. Amber fossils of ancient insects are particularly magnificent. Amber is a transparent rock nodule produced by the petrification of resin that exudes from trees – particularly conifers. Insects and other small animals, such as spiders, can become trapped in the resin where they undergo remarkably little decay. If the amber eventually fossilizes, they remain perfectly preserved.

Trace fossils are not made of the remains of organisms but are the preserved forms of tunnels, worm casts, burrows or tracks made by animals in soft sediments which have turned to rock. They range from the delicate spidery leg imprints made by a moving king crab to the massive footprints of a dinosaur pounding through soft mud. Such trace fossils can reveal much about the lives of the animals that made them. From dinosaur footprints, for example, the gait, stride length and speed of the creature can be deduced.

Where conditions are exceptionally favourable for fossilization, soft body parts as well as hard bones and shells can become fossilized, sometimes producing a remarkable facsimile of the original organism. Particularly famous examples come from the Cambrian Burgess Shale in British Columbia, Canada, and from the Jurassic lithographic limestones of Solenhofen in Bavaria, Germany.

In both instances, animals were rapidly entombed in fine-grained sediments in circumstances that halted microbial decay for long enough to ensure high fidelity preservation. Even the finest legs and antennae can be seen on some of the Burgess Shale arthropods – although they are now half a billion years old. And on the feather vanes of the fossil of the first known bird, *Archaeopteryx*, from Solenhofen even the impressions of tiny individual barbs can be seen.

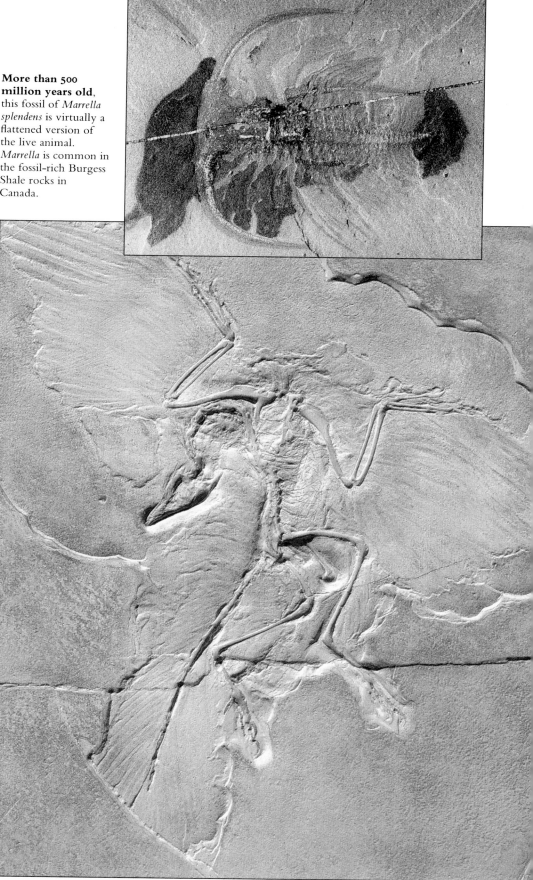

More than 500 million years old, this fossil of *Marrella splendens* is virtually a flattened version of the live animal. *Marrella* is common in the fossil-rich Burgess Shale rocks in Canada.

Almost every bone in the body is visible in this fossil of *Archaeopteryx*, the first known bird.

THE FORMATION OF FOSSILS

Life forms ranging from bacterial cells to huge dinosaur bones have been preserved as fossils. Yet despite their apparent abundance and diversity, fossils provide only an incomplete inventory of Earth's past. If the total fossil record of prehistoric life were a book, it would be one with whole chapters torn out and with many pages, sentences and words completely missing.

Some organisms, such as soft-bodied animals with no shell or skeletal elements, are rarely if ever fossilized. Thus amoebae, ciliates, flatworms, sea slugs, hydras and jellyfish are virtually absent from the fossil record – a serious lack since animals of some of these types are thought to have played key roles in the early phases of animal evolution when multicellular organisms were diversifying. Similarly, plants and fungi with no toughened woody parts do not lend themselves to fossilization. This explains why there are many fossils of the woody parts of plants but few flower fossils.

Equally, for a fossil to have a chance of forming, an organism must die in particular circumstances which allow it to be embedded in sediment and preserved. These sediments – marine muds, sandy or clay deposits in rivers, swamps and lagoons, or blown sand in dune systems of deserts – are the raw material from which sedimentary rocks form. It is in these rocks that most fossils are found.

The chances of fossil formation therefore vary enormously. Fossilization is more likely to happen in marine systems and in lakes and swamps than in most terrestrial habitats. And in grassland or rainforest, fossils are unlikely to form. Almost all the dead animals in such ecosystems are quickly consumed or decomposed, leaving no opportunity for becoming embedded in a sediment.

Even when fossils do form, it is likely that they will never become part of the fossil record available for study. Most sedimentary rocks are ultimately eroded, changed by heat and pressure into metamorphic rock or even melted completely when pushed downward under the crust by the movements of the plates that constitute the Earth's surface. All of these processes destroy fossils. Fossils in sedimentary rocks many miles underground are also unlikely ever to be found. Only those in uneroded sedimentary rocks exposed near the surface are potentially collectable by ordinary methods.

When fossils form in sedimentary rocks, the dead animal or plant is essentially turned into a stone replica of itself. The chemical nature of a fossil goes through different stages and varies between fossil types. A marine mollusc is a typical example. The dead body of the animal rots and its shell becomes engulfed by fine sediment – the beginning of its chemical transformation. The shell is made of crystals of calcium carbonate mixed with a type of tough protein glue synthesized by the mollusc in life. Microorganisms may enter minute pores in the shell and gradually break down the protein within. Minerals also enter the spaces and start to deposit on the crystals making up the shell.

This process consolidates the existing structure and makes it more compact and dense. If the sediments are by this time overlain with others and are beginning the process of compression and chemical change which turns them into rock, similar changes will occur to the developing fossil. It, too, may be compressed and turned into rock like that of the matrix material – the material in which the fossil is embedded.

A calcium carbonate shell may eventually be dissolved by the slightly acidic ground water that percolates through porous rocks. Such dissolution does not necessarily mean the loss of a fossil but can be the first step in another sort of fossil transformation. If the initial fossil leaves a negative imprint in the surrounding rock, this can be filled with a quite different type of rock or mineral. Examples of silica infilling or the transformation of preserved soft parts into iron pyrites can provide beautifully detailed and lasting fossils.

Living ammonite with calcareous shell.

Ammonites, relatives of today's squid, were marine molluscs with hard coiled shells. They were common in the Jurassic period, 200 million years ago.

Dead ammonite lying on sediment on the seabed. Soft body decomposes.

Half a negative impression fossil of ammonite.

Direct transformation fossil of shell. No surface colour remains, but surface sculpturing visible.

Sutures between chambers visible.

Additional layer of sediment above developing fossil.

Once covered by sediment, the outer sheath of the shell decomposed and the outer chamber filled with fine sediment. In time, the shell itself became consolidated by dissolved minerals. A direct fossil formed, as well as two half fossils from the surrounding rock. These show negative impressions of the outer surface.

Acidic conditions remove shell fossil layer, revealing inner mould fossil of shell chambers.

Alternatively, the calcareous shell might have dissolved away, leaving a more resistant internal mould material. On this the inner wavy sutures between the chambers of the coiled shell are visible.

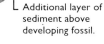

Additional layer of sediment above developing fossil.

Outer chamber of shell filled with sediment. Shell and sediment begin the transformation into rock.

Dissolved minerals added to fossil structure.

Calcareous shell covered by sediment.

After death, the ammonite's soft body soon decomposed. The remaining shell had some closed internal chambers and one open outer chamber, which in life contained the body of the animal.

Translucent calcite (a crystal form of calcium carbonate) has filled the inner chambers of this ammonite fossil (*right*). The outer chamber, which in life housed the animal, must have been filled at an earlier stage by the coarser sediments in which the shell was first embedded.

INTERPRETING THE EVIDENCE

Fossils tell a story about the life of the past. But it is a story that needs careful interpretation. Because they are not always what they seem, fossils often require thorough analysis before all the implications of their form can be properly understood.

The process of interpretation is simplest when the fossil is clearly related at some level to an organism that still exists today. Comparison with a living creature can help "clothe" a skeletal fossil with the non-fossilizable soft parts. In the Pleistocene Turkana beds in east Africa, for example, many fossil shells have been found that are directly related to snails such as *Melanoides* living in tropical fresh waters today.

Ammonites reveal more remote links. These successful and prolific marine cephalopods died out 65 million years ago – at the same time as dinosaurs became extinct. There are many well-preserved fossils of the coiled shells of ammonites but none of the soft body parts. These can be reconstructed with some confidence, however, because the modern *Nautilus* belongs to a group of creatures that were evolutionary precursors of the ammonites. Much can be learned about the ammonites from the bodies of these living animals. Both of them have a many-chambered shell with a siphuncle – a tubular organ for controlling the creature's body density in water. The ammonite probably had a tentacled head for food capture, somewhat like that of *Nautilus*.

With more ancient fossils and ones with no clearly related modern forms, interpretation can be more difficult and tenuous. Over the last 150 years interpretations have been made which now, with the benefit of more knowledge or more explicit fossils, have proved to be erroneous. What was once thought to be the rhinoceros-like horn of the dinosaur *Iguanodon* is now known to have been a large pointed thumb. The full-size replica of *Iguanodon*, constructed in Victorian times and still on display in Crystal Palace Park, London, sports the error to this day.

Hallucigenia, the 550-million-year-old invertebrate fossil from the Cambrian

This simple feather was the first *Archaeopteryx* fossil to be found, in 1860. It closely resembles a wing flight feather of a modern bird, with its vane made up of minute threadlike barbs attached to a feather shaft, or rachis. As in all flight feathers, the rachis is set asymmetrically in the vane.

Air flow over upper surface

Air flow over lower surface

Lift force

Like a modern bird, *Archaeopteryx* had asymmetric feathers – proof that it could fly. The curved aerofoil shape made from the asymmetric feather vane gives lift for flight: the air flow path over the upper surface of the feather is longer than over the lower surface. This produces a pressure difference between the two feather surfaces which creates lift.

Wing bones

Wing bones

Flight muscles

Flight muscles

Shallow keel

Deep keel

Archaeopteryx   Modern bird

Despite its feathered wings, *Archaeopteryx* may have been a poor flier. Its bones were dense and heavy and, unlike modern birds, it did not have a deep keel for the attachment of strong flight muscles.

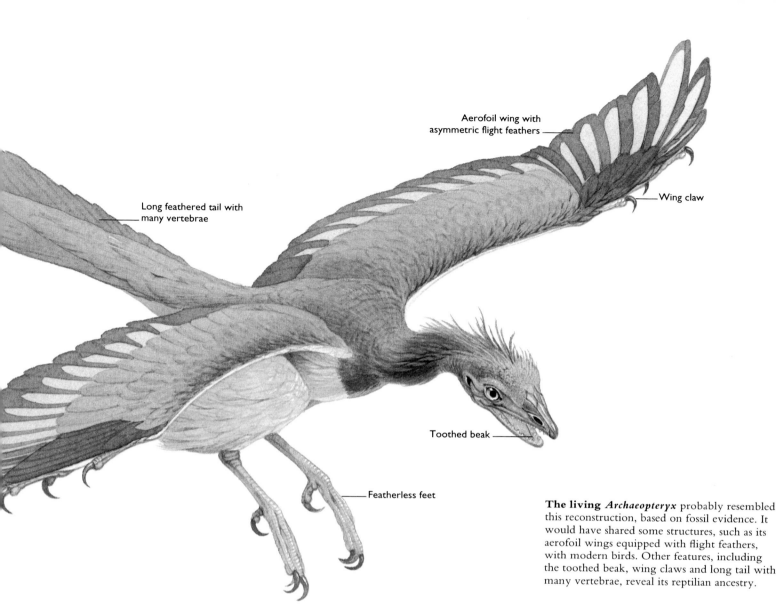

Aerofoil wing with
asymmetric flight feathers

Wing claw

Long feathered tail with
many vertebrae

Toothed beak

Featherless feet

The living *Archaeopteryx* probably resembled
this reconstruction, based on fossil evidence. It
would have shared some structures, such as its
aerofoil wings equipped with flight feathers,
with modern birds. Other features, including
the toothed beak, wing claws and long tail with
many vertebrae, reveal its reptilian ancestry.

Burgess Shale, is another puzzle. When
first described it was thought to belong
to an early animal group which died
out in the Cambrian. Subsequent work
and finds of better fossils from other
parts of the world have revealed that the
initial interpretation had turned the
animal upside down.

What had been thought to be legs
were in fact spines along its back.
Moreover, far from being a member
of a long dead group, *Hallucigenia* and
its relatives are now believed to have
been early forms of the velvet worms
or onycophorans, some of which still
live today (pp. 30–31).

The precise form of fossilized bones
and teeth can often reveal much about
the lifestyle of the original animal. Bones
have flanges, or depressions, which show
the sizes and positions of muscle attach-
ments. If a whole skeleton of fossil bones
is found, these attachment scars help
reconstruct the musculature of the animal and hence its patterns of movement.
Joints between bones often show the
precise ranges of movement available in
a limb or jaw. Teeth have characteristic
arrangements that reveal the diet of their
owner. The teeth of a fish eater, a grinder
of tough plant material or a flesh-eating
carnivore all have particular shapes
suited to their function.

Paleontologists are understandably
excited when fossils are found complete
with fossilized food remains inside

them, which provide explicit information about diet. The fossil of the carnivorous dinosaur *Baryonyx* (recently
discovered in England) had fish scales
inside it that were specifically identified
as belonging to *Lepidotes*, a fish about 3 ft
(1 m) long. *Baryonyx* probably used the
large claws on its forelimbs to hook fish
from the water just as grizzly bears in
Alaska do today.

Among the most famous of all fossils
are those of *Archaeopteryx*, the earliest
known bird. Six fossils have been found,
all from the Solenhofen limestone in
Germany which dates from the late
Jurassic period about 150 million years
ago. From these an attempt can be made
to reconstruct the living creature.

THE LONGEST AGE

Bacteria and simple plants resembling modern blue-green algae were among the first living things on Earth. But their fossilized remains were too small and difficult to recognize to be seen by the earliest scientific fossil hunters of the 18th and 19th centuries who defined the geological time periods in a sequence still used now. The oldest fossil-bearing rock strata that these workers could recognize were those of the Cambrian – named after the Cambrian Mountains in Wales. The Cambrian is now known to have begun about 570 million years ago.

The huge layers of yet older rocks – apparently without fossils – were termed Precambrian, and modern dating techniques have shown that this was indeed "the longest age". The Precambrian began with the earliest phases of the consolidation of the Earth's crust over 4 billion years ago. Sedimentary rocks – the types that can carry fossils –

have been found from around 3.8 billion years ago and through the whole of the rest of the Precambrian

Charles Darwin was troubled by the apparent absence of Precambrian fossils. He knew that the diverse Cambrian life forms were already complex, including creatures such as molluscs and arthropods. He deduced that simpler Precambrian animals and plants must have preceded them but that they must have been soft-bodied, and thus unsuited to fossilization. Darwin was right – the organisms did exist. But some of them did form fossils. In the last few decades new techniques of exposing and interpreting fossils from rock strata all over the globe have revealed a Precambrian world packed with life.

Most Precambrian fossils, some of them 3.5 billion years old or more, are prokaryotes – cells without nuclei. Clusters or, more often, chains of these tiny spheres seemingly resemble today's bac-

teria and blue-green algae in their organization. Presumably they lived in or on marine sediments, some in deep waters, others in shallower seas. There was little or no oxygen in the Earth's atmosphere at this time so these organisms were anaerobic – they operated without oxygen.

On present evidence, it seems that once prokaryotic cells had developed, some 3.8 billion years ago, the pace of evolutionary change was relatively slow for more than 2 billion years. The basic organization of life did not alter, no new environments were colonized and no new ecosystems developed.

The Earth's early atmosphere contained carbon dioxide, nitrogen and small amounts of hydrogen.

The proportion of nitrogen in the atmosphere became greater than that of carbon dioxide during the early Precambrian.

Because it is a light gas, hydrogen was soon lost into space, leaving insignificant amounts in the atmosphere.

Hydrogen

| 4.5 | 4 | 3.5 | 3 |

One agency of change was present, however, in the metabolic strategy of the blue-green algae when their photosynthesis began to produce oxygen. Before they began releasing oxygen as a by-product of sun-driven photosynthesis, there was probably no free oxygen in Earth's atmosphere. Air would have contained nitrogen and carbon dioxide and some hydrogen.

Oxygen from the stromatolite-strewn seas changed the world – but extremely slowly. At first all the oxygen produced was mopped up by rocks and minerals that were in a reduced (oxygen-lacking) chemical condition. Eventually, however, excess oxygen began to accumulate in the atmosphere and had a dramatic effect on living cells – cells that had been trapped in an evolutionary cul-de-sac for 2 billion years. Since oxygen is a dangerous, highly reactive gas, organisms adapted to work in an anaerobic world are at a great disadvantage if the oxygen level increases above a very low level. This initial increase in the amount of atmospheric oxygen seems to have been a profound, mould-breaking stimulus to cellular evolution.

Rising oxygen levels are believed to have triggered the evolution of higher, nucleus-containing, eukaryotic cells. The blue-green algal scum of the Precambrian seas was the metabolic midwife of higher cells – and hence of animals and ourselves.

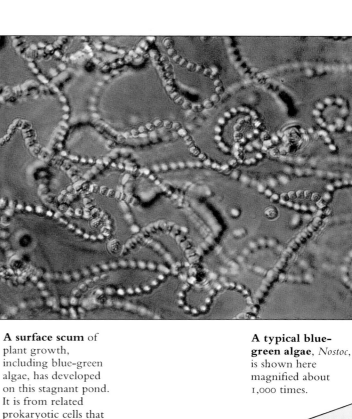

A surface scum of plant growth, including blue-green algae, has developed on this stagnant pond. It is from related prokaryotic cells that all higher forms of life are believed to have evolved.

A typical blue-green algae, *Nostoc*, is shown here magnified about 1,000 times.

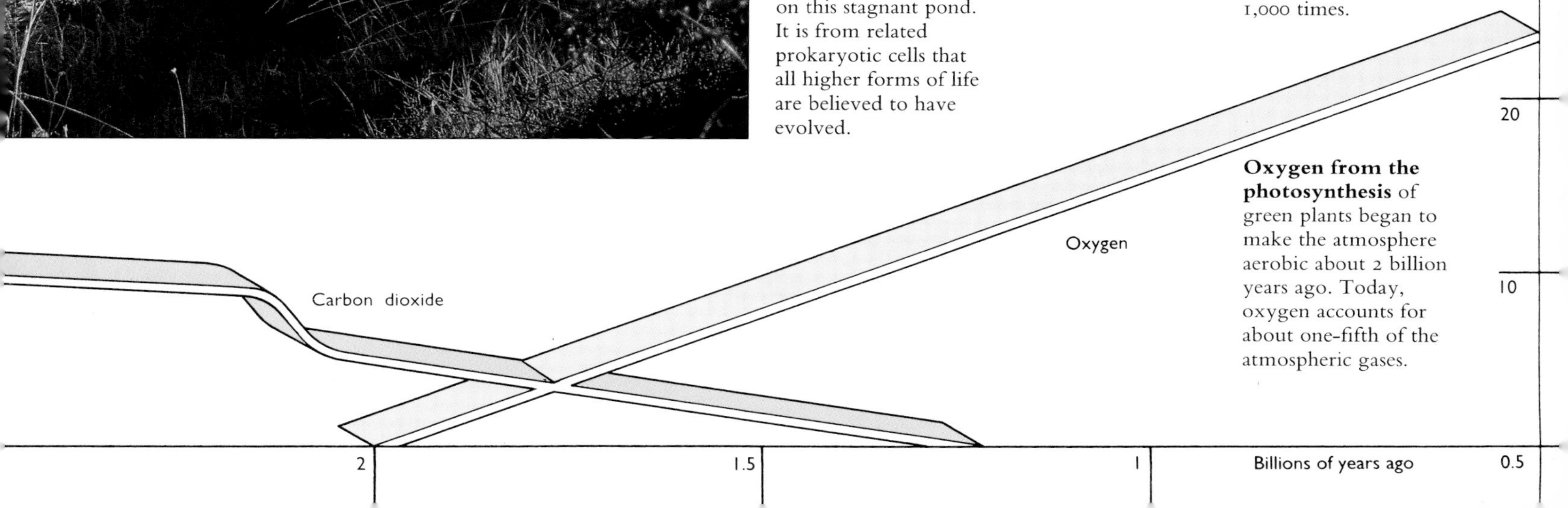

Percentage of atmospheric pressure

100%

90

80

Nitrogen

70

60

50

40

30

20

Oxygen

Oxygen from the photosynthesis of green plants began to make the atmosphere aerobic about 2 billion years ago. Today, oxygen accounts for about one-fifth of the atmospheric gases.

Carbon dioxide

10

2   1.5   1   Billions of years ago   0.5

THE COMING OF COMPLEX CELLS

Oxygen is the key to life both past and present. The evolution of complex, nucleated cells, like those found in modern plants and animals, does appear to have coincided with increased levels of oxygen in the atmosphere. Most evolutionary scientists now agree that increasingly aerobic conditions triggered the development of these eukaryotic cells.

The fact that these cells contained nuclei was only one aspect of their complexity. They also possessed internal subcellular structures – organelles – for carrying out specific tasks. Apart from the organelles known as ribosomes, which are responsible for protein synthesis, all of these organelles were absent from the earlier prokaryotes, that is cells without nuclei.

Perhaps the most important of the organelles developed in cells were the mitochondrion, for aerobic respiration, and the chloroplast, for photosynthesis in aerobic conditions. The mitochondrion can provide energy-rich compounds by using oxygen to "burn" carbon-containing compounds to carbon dioxide and water – a much more efficient form of energy exchange than the anaerobic fermentation of the earlier prokaryotic cells (pp. 56–57).

The only clues to the evolution of such structures as mitochondria and chloroplasts lie worlds apart in the microfossils from a billion years ago and the molecular and biological evidence from cells alive today. The detailed structure of the organelles in eukaryotic cells and the nature of the DNA found in those organelles suggests that these crucial evolutionary acquistions were not simply the result of slow accumulation of random changes – mutations – in the genes of ancestral prokaryotes. Instead, it is possible that some cells took into their own cell contents partner cells of a another kind, cells which had different metabolic abilities.

At first, such combinations were on the basis of intracellular symbiosis – two organisms living intimately together for mutual benefit. But, in time, the genetic and metabolic organizations of host and guest cells became effectively fused and it became impossible to distinguish where one cell began and the other finished.

A scenario of this sort presupposes an ancient prokaryotic host cell with a basically anaerobic metabolism. The cell evolved in a way that changed its genetic organization until it possessed a nucleus with multiple DNA-containing chromosomes (pp. 60–61). It then took into itself a different type of prokaryotic cell, one which had responded to the presence of higher oxygen levels by evolving a process of respiration that used oxygen. Biochemically, this respiration was based on new enzymes and iron-containing molecules termed cytochromes.

Ultimately, this partner cell changed into a mitochondrion, an organelle with

A cell that could engulf others and which had packaged its own genes inside a nucleus may have been the precursor of complex cells. It was into this cell type that other cells moved. Previously independent aerobic bacteria could have become mitochondria. Cells similar to blue-green algae could have become chloroplasts. In both instances the organelles produced came to depend on the nucleus of the host cell to direct some of its metabolic processes. But chloroplasts and mitochondria do retain DNA of their own which codes for some internal proteins. This DNA may be a relic of Precambrian days when the ancestors of mitochondria and chloroplasts were cells in their own right.

To present-day aerobic bacteria

Pre-mitochondrion engulfed

Aerobic bacterium (Pre-mitochondrion)

Nucleated pre-eukaryote

Precursor prokaryote cells

Nucleus forms

Photosynthetic prokaryote (Pre-chloroplast)

the evolutionary jumps were so difficult that they occurred only once or twice in specific cell lineages. Thus, the acquisition of these rare and efficient new cells as partners would confer great benefits on host cells. No other hypothesis explains as well the fact that both mitochondria and chloroplasts contain their own ribosomes and DNA – perhaps a relic from their past as independent cells.

At some point in the late Precambrian non-photosynthetic animal cells developed – perhaps from photosynthetic cells that lost their chloroplasts. Animal eukaryotic cells obtain their nutrients by consuming other cells rather than by constructing them through photosynthesis as plants do. Early single-celled eukaryotes, perhaps similar to today's protozoans, may have engulfed and consumed plant cells.

Still later in the Precambrian came multicellular animals – metazoans. The earliest forms were presumably entirely soft-bodied, like today's flatworms, and no direct fossils have been found up to now. There are, however, some indirect traces of metazoan life from 1 billion to 700 million years ago in the form of fossilized fecal pellets and burrows.

Direct fossils of metazoans have been found from the very last phases of the Precambrian, 680 to 570 million years ago, prime examples being the Ediacaran fossils from the Ediacara Hills in southern Australia. There are two schools of scientific thought about these fossils of animals that lived in the sea and on the seabed 680 million years ago. The first supposes that they are all representatives of present-day groups and their body forms relate to some of today's invertebrates, such as scyphozoan jellyfish, sea pens and annelid worms.

A second view holds that these are chance resemblances and that the Ediacaran animals were early evolutionary experiments in multicellular animal forms, none of which persisted into the Cambrian period.

Whatever the truth, the Ediacaran animals have a special place in the huge cast list of animals that have lived on our planet. They are among the earliest fossils known of distinct animal life.

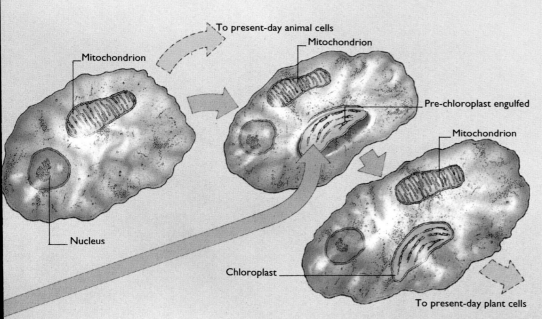

The cells of almost all modern animals have mitochondria but not chloroplasts, while those of plants have both types of organelle. The sequence of evolutionary change shown here suggests that animal cell lines diverged before those of plants, but the reverse could also be true. Animal cells might have evolved from plantlike cells which subsequently lost their chloroplasts.

the same oxygen-using respiration capacity as the free-living cell from which it derived, but now used exclusively for the benefit of the whole composite cell. Similarly, blue-green algal or other photosynthetic cells could have entered cells by the same symbiotic route and transformed into chloroplasts.

This may seem an implausible explanation for evolutionary change. If the ancestral mitochondrial cell could develop aerobic respiration, why could not the host cell have done the same? Perhaps

THE CAMBRIAN EXPLOSION

Life took off in the Cambrian period. By its end, all the main lines of animal types, whose descendants fill the world today, had been established. All of this diversity flourished in the sea; there was still no life on land.

Paleontologists have set the boundary between the Precambrian and the Cambrian at about 570 million years ago. Looking at the rock strata from just before this transition, through the changeover period and into the earliest parts of the early Cambrian period, it is as though a button marked "animal diversity" has been pushed and activated.

Before the transition, there are fossils of the Ediacaran fauna and a few tube-shaped fossils of calcium carbonate such as *Cloudina* and *Sinotubulites*. All except *Sinotubulites* disappear at the transition to be replaced by an ever increasing diversity of fossils of soft- and hard-bodied multi-cellular animals. This sudden expansion in both the range and complexity of multicelled animals has been termed the "Cambrian explosion" and is one of the most extraordinary leaps in the fossil record. By the end of the Cambrian, only 80 million years later, the evolutionary "fall-out" from the explosion had established all the main lines of animal types whose

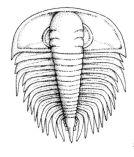

Trilobite

descendants are found in the world today.

Just after the Precambrian/Cambrian transition, the first components of the explosion of life can be seen in the fossil record. There are primitive molluscs of many types, brachiopod lampshells, the first echinoderms, a range of sponge types and some invertebrate forms that are now extinct such as archaeocyathans and trilobites.

The flask-shaped archaeocyathans were strange double-walled creatures, only a fraction of an inch high, with porous skeletons made out of calcium carbonate. Their evolutionary relatives are unknown. Trilobites were among the first arthropods – animals with jointed legs or appendages. They were covered by a shieldlike exoskeleton which they shed many times as they grew.

There are also fossils of individual plates or spines from soft-bodied animals with coverings like those of a minute hedgehog. Spines of another sort comprise the conodont fossils, tiny hooked toothlike shapes made of calcium phosphate. Similar fossils from much later in the fossil record suggest that the conodont-bearing animals had a wormlike form and were perhaps early chordates – the group that gave rise to all vertebrate, or backboned, animals, including humans.

These early Cambrian animals were fossilized because they had mineralized shells. Some scientists believe that the seemingly abrupt appearance of hard-

Archaeocyathan

ened shells was due to the continuing increase in atmospheric oxygen as a result of photosynthesis by the burgeoning diversity of green plants. By 800 million years ago multicellular algal seaweeds had evolved and like their blue-green algal cousins were producing oxygen. Increased oxygen may have made it possible for larger animals to evolve, and with this development came the need for the increased protection afforded by some sort of hard outer shell or exoskeleton.

Marrella

The Burgess Shale fossils from British Columbia in Canada date from the middle Cambrian, around 530 million years ago. They make up one of the most perfectly preserved fossil assemblages of any era. The unparalleled conditions for fossilization have allowed the soft parts of animals as well as mineralized skeletons to be preserved.

The community of animals appears to have lived on seabed sediment at a medium depth of about 500 ft (150 m). The animals were instantaneously engulfed in anaerobic mud by a mudslip from a slope above the seabed. The lack of oxygen in the mud meant that the rate of

Pikaia

EPOCH						
PERIOD				CAMBRIAN	ORDOVICIAN	
ERA		PRECAMBRIAN			PALEOZOIC	

3.8 | 570 | 500 | 440 | 410

BILLIONS OF YEARS AGO | MILLIONS OF YEARS AGO

Earliest known life Bacteria and blue-green algae

First land plants First ferns and seed plants

First invertebrates Invertebrates diversify First trilobites First land invertebrates First insects

First jawless fish First jawed fish

First amphibians

decomposition was slowed, allowing fossils to be formed in the fine-grained shale. The site was discovered early this century, in 1909, by the American paleontologist Charles Walcott and has yielded tens of thousands of fossils.

From the fossil evidence, the Burgess Shale fauna included more than 100 different genera of hard- and soft-bodied animals. There were armoured trilobites and arthropods, such as the many-limbed *Marrella*, many of which are difficult to categorize in modern arthropod groups. Most of these creatures appear to have been bottom scavengers, mud feeders or micropredators.

Sponges and brachiopods abounded as well as the earliest echinoderm sea lilies and a range of molluscs. There was even what is thought to have been a primitive chordate – *Pikaia*. Worms wriggled over the surface of the mud or burrowed into it. *Canadia*, a polychaete annelid worm, is an example of the surface inhabitants while most of the burrowing animals in the seabed, such as *Ottoia*, seemed to have belonged to the now rare priapulid group of worms.

Some of the most exciting Burgess Shale fossils cannot be assigned with certainty to any animal group of today. Steven Jay Gould, author of the book *Wonderful Life*, about the Burgess Shale discoveries, sees this as evidence that these creatures were early evolutionary experiments in animal body forms, only some of which survived the Cambrian period. It is certainly true that few really new animal body plans have arisen since the Cambrian.

Animals such as *Wixwaxia*, *Odontogriphus*, *Opabinia* and the relatively huge, 18-in (45-cm) long, swimming predator *Anomalocaris* have a puzzling mix of characteristics that make Gould's view seem tenable. Looking at this fantastic array of strange creatures, it is easy to believe that at this stage the cards of animal design were being shuffled and dealt over and over again, with few of the hands proving to possess long-term staying power.

WHICH WAY UP?

Work on one Burgess Shale creature is a warning of the dangers of interpreting fossils. *Hallucigenia* was first described in 1977. Named for its "bizarre and dreamlike appearance", this creature seemed the quintessential short-term experiment. It had a single row of tentacles along its back, each tipped with a small claw, and a double row of seven spines on which it walked. Its apparently unique morphology made it hard to guess at its lifestyle.

Recent work with Cambrian fossils from China and a re-examination of the original Burgess species have shown that all these earlier theories about *Hallucigenia* were utterly wrong. Far from being an evolutionary castoff, it seems to have been the direct ancestor of a line that continues to this day. It belongs to a group called the armoured lobopods and is thought to be a forerunner of today's onychophorans or velvet worms which look like a cross between a worm and a centipede.

The tentacles were in fact a set of legs, just like the tubular legs of today's velvet worms. The spines ran along the back and were perhaps a protection against predators. The fossil had been reconstructed upside-down.

Hallucigenia

Armoured lobopod

The time chart below, and on subsequent pages, shows the probable order and timing of some of the major events in the evolution of life on Earth. The main units of geological time are the eras, starting with the Precambrian. Each era is divided into periods and the Cenozoic is further subdivided into epochs. The dates on the chart indicate the time at which each period or epoch is thought to have begun. The branchings of the evolutionary tree in the diagram follow current research, but new fossil discoveries are constantly forcing reappraisals of the timing of such developments.

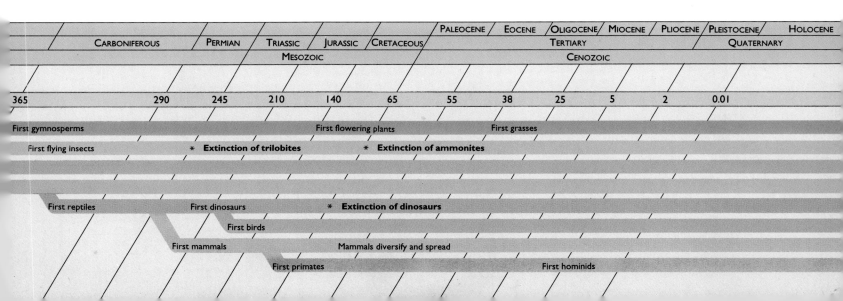

FILLING THE SEAS

Five hundred million years ago there were still only sea-living animals and plants. During the next 140 million years, marine life continued to diversify and such groups as corals, sea urchins and bivalve molluscs appeared. And fish – the first vertebrate animals – began to dominate the seas. Glaciations and major plate movements of the continents affected marine ecosystems throughout this time, which has been divided into three periods defined by their characteristic fossil mixes. In chronological order they are the Ordovician, the Silurian and the Devonian.

Three invertebrate groups that had emerged in the Cambrian now became particularly abundant and varied: the trilobites, the brachiopod lampshells and the graptolites. Extremely common in Ordovician deposits, graptolites were branched colonies of tiny animals which were probably related to an early branch of the chordate line – the hemichordates.

Molluscs, too, were increasing in complexity. The simple, limpet-shaped monoplacophorans began to diversify into bivalves like today's mussels and scallops and into true snail-like gastropods. Corals, echinoderms such as sea lilies and sea urchins, and the colonial bryozoans also evolved.

The very earliest vertebrate fish might have

Sea scorpion

Graptolite

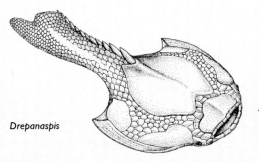

Drepanaspis

begun life in the Cambrian – a few scale fossils have been found from the end of the period. The first complete, direct fossils of fish, though, are Ordovician, with examples such as *Arandaspis* from Australia. This fish had no jaws or teeth, just a round, suckerlike mouth; it did, however, have gills for underwater respiration. By the end of the period there were many more types of these primitive jawless fish in the seas and in the succeeding Silurian they moved into fresh waters. Heavy, slow-moving bottom feeders, they scraped up deposits, including algae, as food, as well as preying on equally slow-moving plankton eaters. Many, such as *Drepanaspis* and *Pteraspis*, had strong bony head shields and bony armour over their bodies. Descendants of the jawless fish – lamprey and hagfish – still survive today.

The first fish with jaws and teeth evolved in the early Silurian, about 80 million years after their jawless ancestors. The development of jaws unleashed a torrent of evolutionary invention. With manipulative jaws and teeth, fish could become active predators, seizing their catches in mid-water and on the bottom.

Like their jawless relatives, the early jawed fish, such as *Climatius*, had skeletons made entirely of gristle or cartilage, not bone. Although fish with bony skeletons soon evolved, the cartilaginous types did not disappear but led to the chondrichthyan group, represented today by sharks, rays and their relatives.

Bony fish, however, became more and

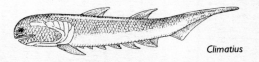

Climatius

more varied in the late Silurian and Devonian, a period often dubbed the age of fishes. They diverged into two types. One, the ray-finned fish (actinopterygians), gave rise to the main groups of modern bony fish. The second type, the fleshy-finned fish (sarcopterygians), had muscular bases to their paired fins. They included the coelacanths and lungfish which survive to this day, as well as the rhipidistians, a group of fish which has unparalleled evolutionary significance: it contained the probable ancestors of land vertebrates.

Nautiloid

Epoch								
Period				Cambrian		Ordovician	Silurian	Devonian
Era		Precambrian				Paleozoic		

3.8
Billions of years ago

570   500   440   410
Millions of years ago

Earliest known life   Bacteria and blue-green algae   First land plants   First ferns and seed plants

First invertebrates Invertebrates diversify First trilobites   First land invertebrates   First insects

First jawless fish First jawed fish

First amphibians

Important developments in the invertebrate world continued through the Silurian and Devonian. Cephalopod molluscs diversified, starting with straight, then coiled, nautiloids. They may have had chambered shells resembling those of today's *Nautilus*. Ammonites derived from nautiloids in the early Devonian and were similar to them in structure. Both groups were expert predators, using their tentacles to catch other invertebrates and small fish in mid-water and on the seabed.

Huge sea scorpions, a now extinct group of arthropods, were 6½ ft (2 m) or more in length and able to capture all but the largest fish. Indeed, the heavy armour-plating of so many early fish may well have been protection against the clawed arms of these carnivores.

By the end of the Silurian simple plants had appeared on land. Plant ecosystems were soon colonized by the first land animals, probably scorpions and millipedes. In the Devonian, plant life grew more complex and tree ferns and giant horsetails created the first forests. And on to this land came the amphibians.

A powerful predator, *Dunkleosteus* was a late Devonian fish, at least 11ft (3.3 m) long. Sinuous movements of its strong body and eel-like tail would have swept this great creature through the sea in search of its fish prey.

Coelacanth

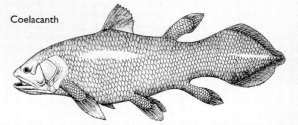

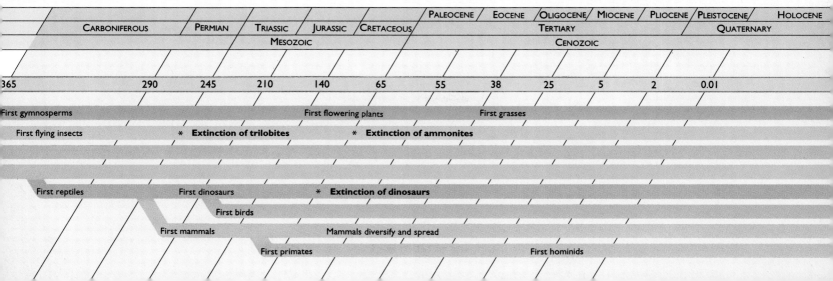

						PALEOCENE	EOCENE	OLIGOCENE	MIOCENE	PLIOCENE	PLEISTOCENE	HOLOCENE
	CARBONIFEROUS	PERMIAN	TRIASSIC	JURASSIC	CRETACEOUS				TERTIARY		QUATERNARY	
			MESOZOIC					CENOZOIC				
365	290	245	210	140	65	55	38	25	5	2	0.01	

First gymnosperms   First flowering plants   First grasses

First flying insects   * **Extinction of trilobites**   * **Extinction of ammonites**

First reptiles   First dinosaurs   * **Extinction of dinosaurs**

First birds

First mammals   Mammals diversify and spread

First primates   First hominids

THE MOVE ON TO LAND

Springtail

The moment when animal life first moved on to land is one of the great landmarks of evolution. From the beginnings of cellular life on Earth, about 4 billion years ago, organisms had lived only in the sea. But by the Silurian period, about 400 million years ago, there was some life in fresh waters. Around the same time, plants began to colonize the muddy fringes of coastal swamps and became more complex than the simple frondlike seaweeds. These new plant forms were marked by the fact that they had within them systems of minute tubes or vessels — they were vascular. As in modern plants, these vessels carried water and nutrients from the soil and the products of photosynthesis round the plant.

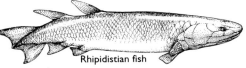

Dragonfly

With this transport system, plants were endowed with the possibilities for becoming more complex anatomically, with parts designed for different functions — leaves for photosynthesis, stems for support, and roots for anchorage and absorbing water and minerals. This was an evolutionary landmark, a "design feature" that opened up terrestrial habitats to

- Spiderlike arachnid

plants. By the Devonian, giant horsetail plants and huge tree ferns were growing on land — albeit in wet swampy areas.

Once these photosynthesizing plants were established on land, the potential was there for more complex organisms, all dependent on the productivity of plants — there was food to eat on land. The first animals to evolve specializations that allowed them to fill niches in this new plant world seem to have been arthropods — their remains have been fossilized among swamp plants. These primitive wingless insects, such as springtails and bristletails, probably fed on plant remains, algal cells and on the fungi that decomposed dead plants. Predators, perhaps spiderlike arachnids, then evolved that fed on the plant eaters. Worms and other soft-bodied invertebrates probably lived in the swamps too, but did not fossilize well.

Rhipidistian fish

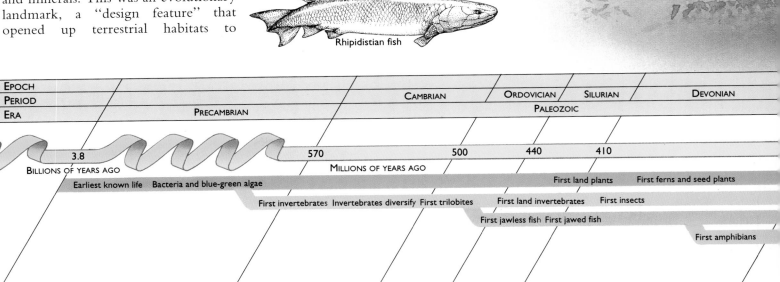

EPOCH								
PERIOD				CAMBRIAN		ORDOVICIAN	SILURIAN	DEVONIAN
ERA		PRECAMBRIAN				PALEOZOIC		

3.8
BILLIONS OF YEARS AGO

570 500 440 410
MILLIONS OF YEARS AGO

Earliest known life Bacteria and blue-green algae

First land plants First ferns and seed plants

First invertebrates Invertebrates diversify First trilobites First land invertebrates First insects

First jawless fish First jawed fish

First amphibians

Throughout the late Devonian and Carboniferous periods land plants such as club mosses became increasingly diverse and began to grow in drier, fully terrestrial environments as well as swamps. More types of invertebrate, including scorpions and dragonflies, spread through the different vegetation systems, probably living in much the same way as they do today.

The first vertebrate animals moved on to land during the transition between the Devonian and the Carboniferous periods, about 370 million years ago. Late Devonian deposits from warm swamp forests that grew where glaciers cover Greenland today contain the fossils of amphibians, the first land vertebrates.

Amphibians are believed to have derived from the rhipidistians, a group of fleshy-finned sarcopterygian fish. These carnivorous fish lived in salt and fresh waters, had muscular paired fins and possessed lungs or organs like them. Some 350 million years ago, they evolved into air-breathing creatures equipped with four limbs that could walk on land. Like amphibians today, their eggs and tadpoles remained aquatic and their lives were divided between water and land.

From these first amphibians such as *Ichthyostega*, which still had a fishlike tail and scales, a range of types soon evolved. Some even resembled today's crocodiles in their size and shape. Their most successful developmental line was the labyrinthodont group, which included forms as small as the modern newt and others, such as *Eogyrinus*, which were up to 15 ft (4.5 m) long.

Throughout the second half of the Carboniferous and the first half of the Permian, the land was dominated by a wide range of carnivorous amphibians. The more aquatic forms fed on fish, the more land-based forms on insects, other invertebrates and smaller amphibians. They may also have preyed on the early reptiles which had evolved from one group of amphibians by about 300 million years ago, in the Carboniferous.

Today there are few groups of amphibians and all are small creatures such as frogs, toads and newts. But 300 million years ago almost every large land animal was an amphibian.

Large amphibians such as *Eryops* were among the first vertebrates to live on land. *Eryops* grew up to 6½ ft (2m) long. Its thickset body was covered with strong bony plates which may have helped support it on land.

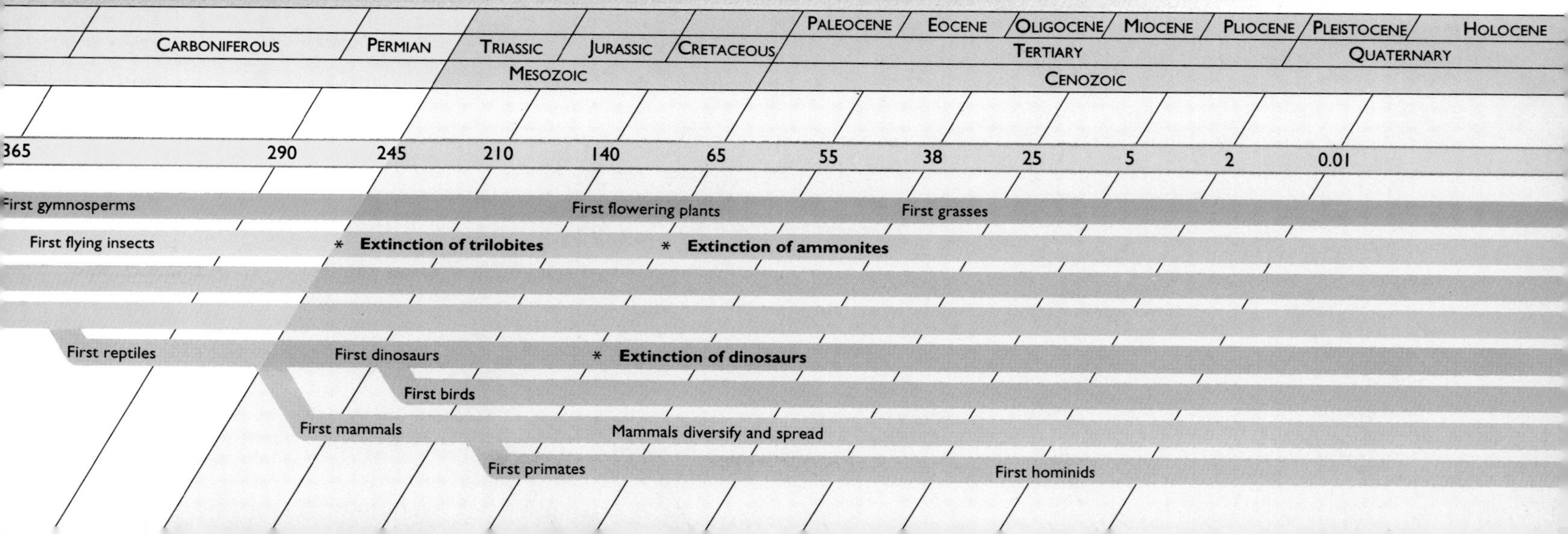

			PALEOCENE	EOCENE	OLIGOCENE	MIOCENE	PLIOCENE	PLEISTOCENE	HOLOCENE
CARBONIFEROUS	PERMIAN	TRIASSIC	JURASSIC	CRETACEOUS			TERTIARY		QUATERNARY	
		MESOZOIC					CENOZOIC		
365	290	245	210	140	65	55	38	25	5	2	0.01

First gymnosperms

First flowering plants

First grasses

First flying insects

* **Extinction of trilobites**

* **Extinction of ammonites**

First reptiles

First dinosaurs

* **Extinction of dinosaurs**

First birds

First mammals

Mammals diversify and spread

First primates

First hominids

THE AGE OF DINOSAURS

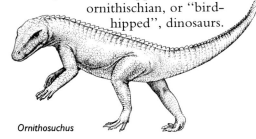

Pterosaur

Brachiosaurus

Ornithosuchus

Crocodile

Reptiles were the first truly terrestrial animals. They dominated life on land through most of the Triassic, Jurassic and Cretaceous periods, from 245 to 65 million years ago. There were many types of reptile but dinosaurs were the most successful – they reached a pinnacle of evolutionary diversity.

Amphibians, from which the reptiles evolved, were never more than partially terrestrial. Their reproductive biology kept them dependent on water: both their eggs and tadpoles were completely aquatic. The evolutionary coup of the reptile group was the development of a shelled egg that did not dry out in air. With eggs of this type, some reptiles could cut the ecological "umbilical cord" that linked their amphibian relatives to water and live entirely on land.

The first reptiles to make this adjustment evolved in the late Carboniferous, about 300 million years ago. *Hylonomus* was probably typical of these. A small reptile – about 8 in (20 cm) long – it fed on insects in the giant club moss forests of the time.

Through the following Permian and Triassic ages evolutionary offshoots of the earlier primitive reptiles colonized high, dry land, including deserts. Their drought-resistant eggs, dry, evaporation-reducing scaly skin and system

Hylonomus

of internal fertilization meant that they could prosper in such arid habitats, where no previous vertebrate could have lived. Some of the reptiles that developed then, such as tortoises and turtles, still survive today. At the same time as reptiles were beginning to thrive on land, large, streamlined marine forms, such as plesiosaurs and ichthyosaurs, flourished in the sea.

Dinosaurs and their near relatives, crocodiles and the flying pterosaurs, were the "ruling reptiles" or archosaurs. Today, only the crocodilians remain – feeble remnants of the mighty zenith of reptilian evolution.

That evolution began at the end of the Permian, about 250 million years ago. A new group of reptiles – the eosuchians or "crocodiles of the dawn" – arose, which subsequently diversified into an extremely successful group of animals known as thecodonts or "socket-toothed" reptiles. Most were large carnivores that moved on four legs in the normal reptilian way – *Erythrosuchus* was a typical example. But thecodonts also developed that could walk on hind legs alone, such as *Ornithosuchus*, and from these reptiles evolved the dinosaurs. Crocodiles and winged pterosaurs came from other thecodont stocks, while from separate eosuchian ancestors came the lines that produced lizards and snakes.

But it was the dinosaurs that ruled the age. From about 200 million years ago, soon after they evolved from thecodonts, until their ultimate demise at the end of the Cretaceous 65 million years

ago, dinosaurs filled almost every available niche for large vertebrate land animals.

There were carnivorous, herbivorous and omnivorous dinosaurs. They ranged from about the size of a household cat to the biggest land animals that have ever lived.

Two major groups of dinosaurs developed, and they can be distinguished anatomically by their hip bones: the saurischian, or "lizard-hipped", and the ornithischian, or "bird-hipped", dinosaurs.

Among the first group are *Tyrannosaurus*, probably the largest carnivore that has ever lived, and smaller meat eaters, most of which ran on two legs. It is from these small carnivorous dinosaurs that the first bird, *Archaeopteryx*, is thought to have evolved in the late Jurassic (pp. 24–25). Also in this group were huge herbivores such as *Diplodocus*

EPOCH								
PERIOD					CAMBRIAN	ORDOVICIAN	SILURIAN	DEVONIAN
ERA		PRECAMBRIAN				PALEOZOIC		

3.8

570   500   440   410

BILLIONS OF YEARS AGO   MILLIONS OF YEARS AGO

Earliest known life   Bacteria and blue-green algae

First land plants   First ferns and seed plants

First invertebrates   Invertebrates diversify   First trilobites   First land invertebrates   First insects

First jawless fish   First jawed fish

First amphibians

Protoceratops **defends** its nest from a marauding *Oviraptor* that is trying to steal its eggs. Both dinosaurs lived in Asia during the Cretaceous. *Protoceratops* was a heavily built plant eater while *Oviraptor* was a fast-moving carnivore.

and *Brachiosaurus*. These vast plant eaters were the largest land animals ever to walk on Earth.

Most of the ornithischian dinosaurs were herbivores that took advantage of the great expansion in flowering plant diversity during the Cretaceous. Their jaws and teeth were highly specialized for the efficient processing of leaves. This group includes such dinosaurs as

Iguanodon, Stegosaurus and *Triceratops* as well as the duck-billed hadrosaurs.

At the end of the Cretaceous, 65 million years ago, there was an abrupt change. Dinosaurs,

Diplodocus

pterosaurs and the large marine reptiles all became extinct. There have been many theories about this mass extinction (pp. 186–87), but whatever its cause it ended the longest-running land vertebrate "star role" ever. No other group, has dominated Earth for as long as dinosaurs.

	CARBONIFEROUS	PERMIAN	TRIASSIC	JURASSIC	CRETACEOUS	PALEOCENE	EOCENE	OLIGOCENE	MIOCENE	PLIOCENE	PLEISTOCENE	HOLOCENE
								TERTIARY			QUATERNARY	
			MESOZOIC					CENOZOIC				
65		290	245	210	140	65	55	38	25	5	2	0.01

First gymnosperms   First flowering plants   First grasses

First flying insects   * **Extinction of trilobites**   * **Extinction of ammonites**

First reptiles   First dinosaurs   * **Extinction of dinosaurs**

First birds

First mammals   Mammals diversify and spread

First primates   First hominids

THE EVOLUTION OF LAND PLANTS

Marine alga

Plants originated in the sea around one billion years ago. Their subsequent evolution, as revealed by the testimony of fossils and the groups of plants that dominate our world today, reached a dramatic climax when they conquered the multitude of habitats on land.

In the surface waters of the Precambrian seas the photosynthetic machinery of blue-green algae became incorporated into the earliest eukaryotic algal cells, probably by a series of symbiotic unions (pp. 28–29) By the Cambrian period, these ancestral eukaryotes had evolved into a wide range of single-celled and multicellular algae. Their descendants still thrive in freshwater and marine environments. They include simple green algae, such as the sea lettuce (*Ulva*), red and brown "seaweeds" and single-celled diatoms, dinoflagellates and euglenoids.

By around 400 million years ago, the previously bare land surfaces of the planet were inhabited by at least two plant types that were not entirely aquatic. One, typified by the fossil *Sporogonites*, had primitive rootlike structures, called rhizoids, for absorbing water from the land on which it grew. The tips of upright stalks on its upper surface carried spores which, when dispersed, grew into new plants.

The second type, exemplified by the fossil *Cooksonia*, had evolved xylem vessels – hollow tubes made from cellulose and lignin (pp. 94–95). These vessels provided both an internal support system that allowed the plant to grow vertically and a means for transporting water absorbed from the soil to all parts of the plant. Furthermore, a thickened layer of cells covering the whole plant cut down water loss by evaporation. These adaptations opened up the evolutionary options for a staggering diversity of truly terrestrial plants, such as club mosses, horsetails, ferns, cycads, conifers and the flowering plants.

During the Devonian period, ferns and horsetails emerged as separate groups and some, by the Carboniferous, had produced treelike

Club moss

Seed fern

species. *Lepidodendron* was the largest of the club mosses, reaching 150 ft (45 m) above the ground. These spore-producing plant groups shared a complex cycle of reproduction (see box) involving the alternation of two generations – a diploid sporophyte and a haploid gametophyte. When seed ferns and gymnosperms, such as conifers, evolved they combined this two-generation cycle into one plant. The gamete-producing haploid generation was retained as a reproductive organ on the larger diploid plant. As a result the gymnosperms were the first plants to produce seeds rather than spores. Wind-blown pollen from male organs fertilized ovules which became seeds. The seeds of seed ferns, cycads, ginkgo trees and conifers, such as the monkey puzzle tree, evolved water-proofed outer cases and an internal food store. These adaptations meant that seeds were tougher than spores and enabled them to survive periods of drought before germinating into new plants.

Flowering plants, or angiosperms, were equipped with these reproductive strategies when they evolved during the Cretaceous period. By the start of the Tertiary, the success of their flowers as reproductive organs meant that angiosperms such as

Gymnosperm (monkey puzzle tree)

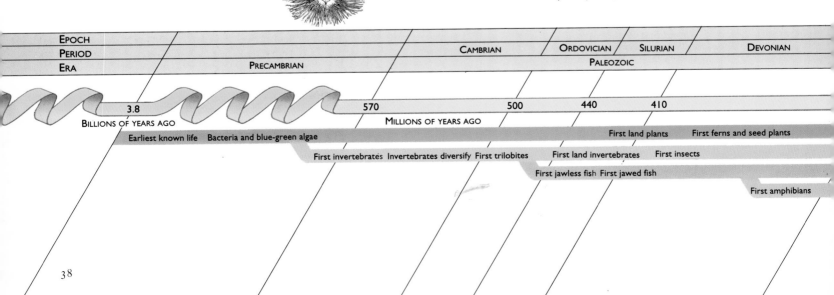

Epoch							
Period				Cambrian	Ordovician	Silurian	Devonian
Era	Precambrian				Paleozoic		

3.8   570   500   440   410

BILLIONS OF YEARS AGO   MILLIONS OF YEARS AGO

Earliest known life   Bacteria and blue-green algae   First land plants   First ferns and seed plants

First invertebrates   Invertebrates diversify   First trilobites   First land invertebrates   First insects

First jawless fish   First jawed fish

First amphibians

Flowering plant (magnolia)

magnolias domi-
nated plant com-
munities. Both
the sepals and
petals of an angio-
sperm's flower
probably evolved
from adapted leaves. The ovules became
completely enclosed in a protective surround, or carpel. For pollination to
succeed, pollen grains had to land on a
stigma – the sticky extension of a carpel –
and grow a pollen tube through the style
to fertilize the ovule. The seeds thus
produced had protective coats and were
often contained in fruits that were
attractive to animals – once the fruits
were eaten the seeds would be dispersed
in the animals' feces.

Grass

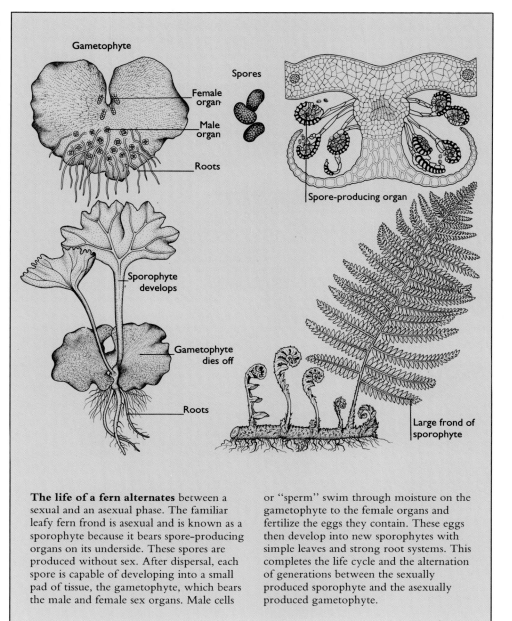

Gametophyte

Spores

Female organ

Male organ

Roots

Spore-producing organ

Sporophyte develops

Gametophyte dies off

Roots

Large frond of sporophyte

The life of a fern alternates between a
sexual and an asexual phase. The familiar
leafy fern frond is asexual and is known as a
sporophyte because it bears spore-producing
organs on its underside. These spores are
produced without sex. After dispersal, each
spore is capable of developing into a small
pad of tissue, the gametophyte, which bears
the male and female sex organs. Male cells
or "sperm" swim through moisture on the
gametophyte to the female organs and
fertilize the eggs they contain. These eggs
then develop into new sporophytes with
simple leaves and strong root systems. This
completes the life cycle and the alternation
of generations between the sexually
produced sporophyte and the asexually
produced gametophyte.

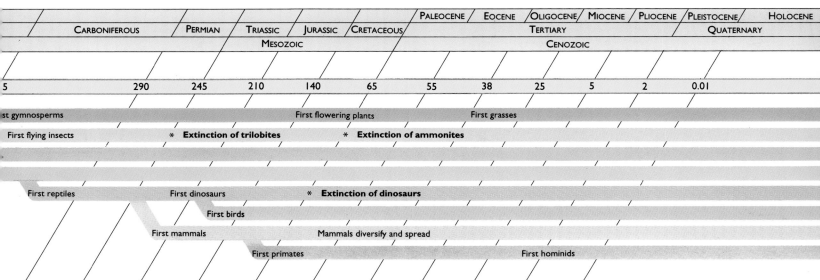

			PALEOCENE	EOCENE	OLIGOCENE	MIOCENE	PLIOCENE	PLEISTOCENE	HOLOCENE
CARBONIFEROUS	PERMIAN	TRIASSIC	JURASSIC	CRETACEOUS			TERTIARY		QUATERNARY
		MESOZOIC					CENOZOIC		

| 5 | 290 | 245 | 210 | 140 | 65 | 55 | 38 | 25 | 5 | 2 | 0.01 |

st gymnosperms   First flowering plants   First grasses

First flying insects   * **Extinction of trilobites**   * **Extinction of ammonites**

First reptiles   First dinosaurs   * **Extinction of dinosaurs**

First birds

First mammals   Mammals diversify and spread

First primates   First hominids

THE MAMMAL TAKEOVER

About 65 million years ago a global catastrophe changed the course of evolution. All the dinosaurs, pterosaurs and large marine reptiles died out and the age of mammals was ushered in. The success story of the mammals clearly shows the power of evolution to produce a huge range of specialized versions of a basic body plan, once given the chance. Evidence for the cause of this major

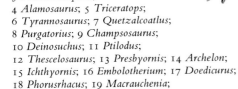

1 *Pachycephalosaurus*; 2 *Pteranodon*;
3 *Parasaurolophus*;
4 *Alamosaurus*; 5 *Triceratops*;
6 *Tyrannosaurus*; 7 *Quetzalcoatlus*;
8 *Purgatorius*; 9 *Champsosaurus*;
10 *Deinosuchus*; 11 *Ptilodus*;
12 *Thescelosaurus*; 13 *Presbyornis*; 14 *Archelon*;
15 *Ichthyornis*; 16 *Embolotherium*; 17 *Doedicurus*;
18 *Phorusrhacus*; 19 *Macrauchenia*;
20 *Megatherium*; 21 *Thylacosmilus*; 22 *Hyaenodon*;
23 *Arsinotherium*; 24 *Megaceros*; 25 *Elephant*;
26 Rhinoceros; 27 Ostrich; 28 Springbok; 29 Bald eagle;
30 Bison; 31 Fruit bat; 32 Crocodile; 33 Giraffe;
34 Chimpanzee; 35 Tiger; 36 Kangaroo.

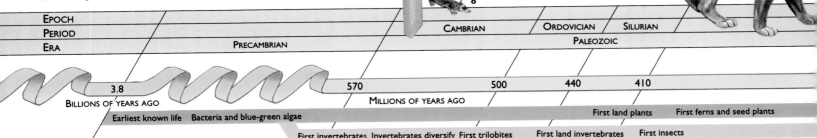

EPOCH						
PERIOD				CAMBRIAN	ORDOVICIAN	SILURIAN
ERA		PRECAMBRIAN			PALEOZOIC	

3.8

BILLIONS OF YEARS AGO

570   500   440   410

MILLIONS OF YEARS AGO

Earliest known life   Bacteria and blue-green algae

First land plants   First ferns and seed plants

First invertebrates   Invertebrates diversify   First trilobites   First land invertebrates   First insects

First jawless fish   First jawed fish

First amphibians

upheaval in vertebrate evolution is complex, but there is a well-supported theory that the impact of a huge object or series of objects from space created a massive disturbance of the Earth's climate (pp. 184–85). Plants died, and with them the plant eaters and their predators. Dinosaurs were unable to survive the catastrophe.

Small mammals had lived with dinosaurs for millions of years. They first emerged at almost exactly the same time as the dinosaurs, about 200 million years ago, but there were few vacant niches for them to fill. Land habitats were saturated with dinosaurs.

The origins of mammals lie back in the early history of reptilian development, long before any dinosaurs walked the Earth. Their reptilian ancestors, such as therapsids which existed from the late Carboniferous onward, shared with mammals a particular type of skull. A single large opening in the wall of the skull behind the eye socket allowed highly efficient jaw-closing muscles to develop, greatly increasing the power of the jaws. By the early Triassic, some therapsids, such as *Cynognathus*, looked very like mammals and may even have developed body hair.

By the late Triassic, about 190 million years ago, early mammals had evolved from therapsids and lived alongside dinosaurs. These first mammals were probably small, insignificant shrewlike creatures, living in dense vegetation, under rocks or in deep burrows and feeding mainly on insects or other small invertebrates. Because they were warm-blooded they could move about and feed in the cool of the night when many of the cold-blooded dinosaurs were perhaps less active.

When the catastrophe came, these small animals were, it appears, less affected than the more specialized dinosaurs. With the demise of those dinosaurs, opportunities for evolution were suddenly available. The specialization potential of mammals was released, allowing them to evolve into the huge range of creatures known today.

				PALEOCENE	EOCENE	OLIGOCENE	MIOCENE	PLIOCENE	PLEISTOCENE	HOLOCENE
		JURASSIC	CRETACEOUS			TERTIARY			QUATERNARY	
						CENOZOIC				

| 5 | 290 | 245 | 210 | 140 | 65 | 55 | 38 | 25 | 5 | 2 | 0.01 |

First gymnosperms   First flowering plants   First grasses

First flying insects   * **Extinction of trilobites**   * **Extinction of ammonites**

First reptiles   First dinosaurs   * **Extinction of dinosaurs**

First birds

First mammals   Mammals diversify and spread

First primates   First hominids

OUR HUMAN ORIGINS

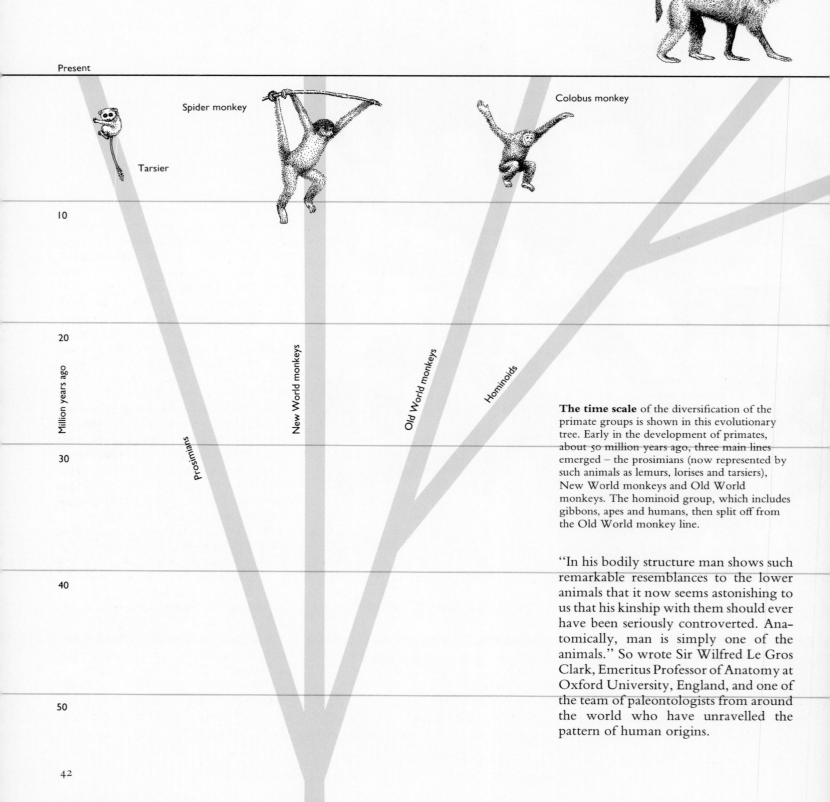

Baboon

Spider monkey

Colobus monkey

Tarsier

Present

10

20

Million years ago

30

40

50

Prosimians

New World monkeys

Old World monkeys

Hominoids

The time scale of the diversification of the primate groups is shown in this evolutionary tree. Early in the development of primates, about 50 million years ago, three main lines emerged – the prosimians (now represented by such animals as lemurs, lorises and tarsiers), New World monkeys and Old World monkeys. The hominoid group, which includes gibbons, apes and humans, then split off from the Old World monkey line.

"In his bodily structure man shows such remarkable resemblances to the lower animals that it now seems astonishing to us that his kinship with them should ever have been seriously controverted. Anatomically, man is simply one of the animals." So wrote Sir Wilfred Le Gros Clark, Emeritus Professor of Anatomy at Oxford University, England, and one of the team of paleontologists from around the world who have unravelled the pattern of human origins.

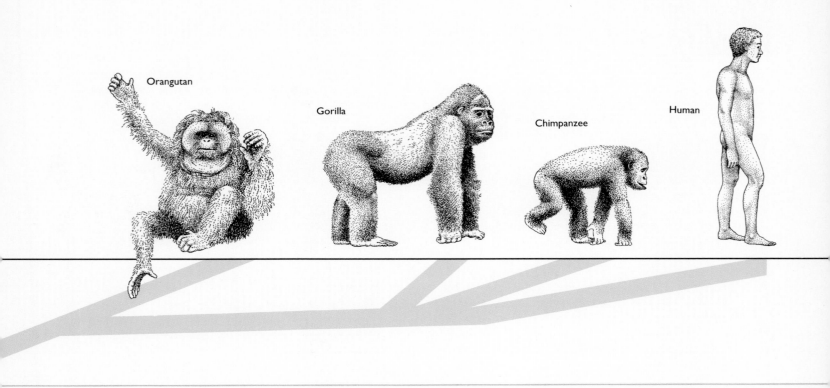

Orangutan

Gorilla

Chimpanzee

Human

Homo sapiens is a mammal and one of 185 living species in the primate order. It is the evolution of other primates that is the key to understanding the ancestry of the human species.

Primates today are split into four major divisions: the prosimians, such as lemurs, tarsiers, lorises and bushbabies; New World monkeys, for example marmosets, howler and spider monkeys; Old World monkeys, such as baboons and macaques; and the hominoid primates, which include the apes and humans. With a few exceptions, primates live in the tropics and subtropics, usually in forested areas. By origin and in their general physical organization, primates seem to be arboreal creatures.

Many aspects of primate anatomy and function can be thought of as evolutionary adaptations for life in the trees. Hands and feet have five sensitive, mobile fingers or toes, which can firmly grasp branches. Teeth and the organization of the gut are suited to plant eating. Perhaps because of their need for accurate, agile movement in the forest canopy, primates have excellent eyesight. Most have short snouts and forward-facing eyes, providing high-precision, stereoscopic vision. Compared to other mammals, primates have upgraded vision and downgraded their sense of smell. The brains of higher primates, such as apes, are comparatively

bigger and more complex than those of other mammals. This, too, may be linked with the need to coordinate rapid movement through the trees.

Nearly all primates have a highly structured family life. They produce small numbers of offspring at a time – often only one per litter – and suckle for a long period. There are obvious adaptive benefits to such small litters: it would be hard to carry and feed a large number of youngsters while living in the trees. Primates are also extremely social animals. Their keen visual abilities and big brains may have pre-adapted them for the complex signalling, communication and group organization that take place in their societies.

Primates probably evolved from simpler shrewlike ancestors toward the end of the Cretaceous, about 70 million years ago, in Europe or North America. The transition from creatures similar to tree shrews to the first true primates took place in landscapes still dominated by dinosaurs. With the evolution of flowering plants there were more trees than ever before, but dinosaurs did not live in trees so there were vacant niches available for these early mammals to fill.

Alternative theories on the evolution of primates dispute their links with the trees, however, and place the early species on the ground. They suggest that

the manipulative hands and forward-facing eyes were adaptations for locating and seizing insect prey, not for clambering around in trees.

From around 60 million years ago, after the demise of the dinosaurs, the early prosimian primates, similar to lemurs and tarsiers, diversified rapidly. The more advanced anthropoid (human-like) primates – monkeys and apes – evolved about 50 million years ago. They are distinguished from other primates by features such as a large, rounded cranium (holding a big brain), only two nipples on the chest, and flat nails instead of claws on fingers and toes. Some types walk or run on all fours, others use their powerful arms to swing expertly from tree to tree. Many can sit up, freeing their hands for the manipulation of food.

The New and Old World monkeys have many similarities because they probably evolved from groups of the same type of tarsier-like primates. The New World monkeys, though, have well-separated, sideways-facing nostrils and many have prehensile tails that can support their weight. Old World forms have close-set, usually downward-pointing nostrils and non-prehensile tails.

Apes evolved from the Old World group and it is on this most recent branch of the primate "bush" of species that *Homo sapiens* belongs.

Fossil remains of monkeys and apes are rare finds indeed. Primates have probably always lived, as they do now, in tropical areas where there is little sediment production and where predators and decomposition destroy most carcasses. With the help of the fossils that have been found, however, the broad outlines of modern human origins can be drawn.

Today, monkey species greatly outnumber apes. But 20 million years ago the situation was the reverse – apes were numerically dominant and as diverse in appearance and lifestyle as monkeys are today. These ape species originally lived only in Africa, but 18 million years ago, when Africa rejoined Eurasia, advanced primates were able to move across the connecting land bridge. In the fossils produced soon after this time are creatures that some paleontologists feel might have been on a line or lines of development leading to modern humans.

From the 1960s until the 1980s the early ape *Ramapithecus* from Europe and Asia, which first appeared about 15 million years ago, was believed to be such an animal. Its short face, heavy jaws

and enamelled teeth were thought to suggest hominid affinities. However, recent DNA evidence has shown that *Ramapithecus* is unlikely to have been a direct human ancestor. It suggests that the great ape-human divergence happened more recently, between five and eight million years ago, so *Ramapithecus* was too early to be a genuine link. A much better candidate for the transitional animal is now thought to be *Australopithecus afarensis*. This creature, which looked like a small, upright chimpanzee, lived about four million years ago and may have been a direct ape-man ancestor of humans.

Studies of the DNA of humans and modern apes show that gorillas and chimpanzees are our closest living relatives. The DNA of a chimpanzee, for example, is about 98 percent identical to that of a human. Careful comparison of the human skeleton with that of a gorilla or chimpanzee reveals that humans have ape bodies adapted for bipedalism – walking on two legs. The lower portion of the human body has evolved to facilitate the load-bearing and balance

requirements of walking upright. A crucial element of this adaptation pattern is the form of the pelvis and the angle of the femur, or thigh bone, with the upper part of the knee.

In apes, the femur and lower leg are in a straight line. In humans they are at an angle, which enables both feet to be under the centre of gravity of the body when walking upright. Intriguingly, *Australopithecus afarensis* had legs in the human pattern, suggesting it was bipedal most of the time. Other human bipedal specializations include the simplification of the foot from a branch-grasping structure with an opposable big toe to a simple flat flexible platform for walking.

At walking speed, bipedalism is a more energy-efficient method of locomotion than movement on four legs. It may have evolved as an adaptation which made the wide-ranging, omnivorous ape-man an even more successful forager. But it had consequences far beyond this. Standing, walking and running on two legs freed up the forelimbs and hands to an unprecedented extent for delicate manipulation. In a

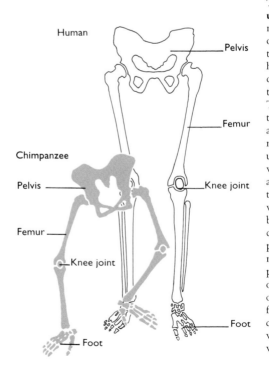

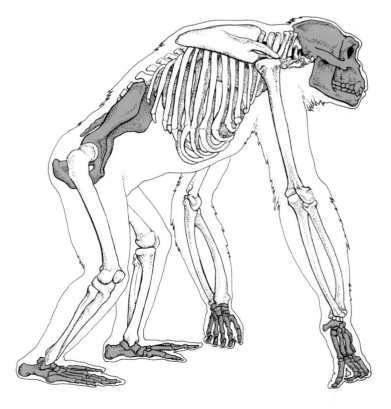

The ability to walk upright habitually might have been one of the earliest traits that marked out the human line of development from that of other apes. The flattened pelvis in the human provides attachment for thigh muscles which give upright stability, while the femurs are angled inward so that the lower limbs are vertical under the body's weight. In the chimpanzee the higher pelvis has most muscles at the rear to power movement on all fours. The outward directed femurs allow only a crude waddling gait when a chimpanzee walks upright.

primate which already possessed adaptations such as precise coordination, social structures, good eyesight and a big brain, this evolutionary opportunity led to a great expansion in tool use. Other living great apes, such as chimpanzees, use tools but in very restricted ways. In australopithecines and then humans, this type of behaviour became more varied and more important. Tool use became part of almost every human activity.

By the time the first members of our genus, *Homo*, had evolved from the African australopithecines about 2 to 1.5 million years ago, another major development had occurred. The fossil skulls of forms such as *Homo habilis* ("handy man"), a maker of simple stone tools, reveal the first sign of this. Although the upper cavity of these skulls suggests a brain volume of only half that of a modern human, casts of this cavity show a characteristic bulge. This corresponds with Broca's area – a zone of the cerebral cortex now known to be responsible for speech production – and suggests that these early humans had the power of speech. An upright ape that made and used tools and could talk was in the world. Real humans had arrived.

Comparison of the skeletons of a modern human and a gorilla show a range of evolutionary changes. Areas of particular importance are highlighted. One immediate contrast is that the human has much longer legs than arms – necessary for upright movement. The gorilla, which usually climbs and moves on all fours, has longer arms than legs.

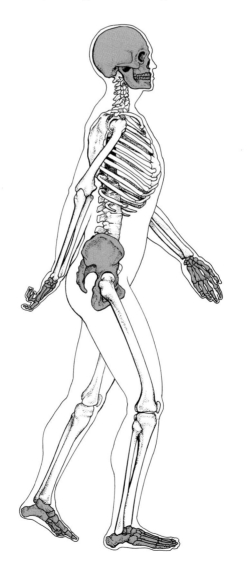

SKULL EVOLUTION

The chimpanzee is perhaps the nearest living ape relative of humans, and interesting comparisons can be drawn between its skull and those of early and modern humans. The ape skull has a ridge on top for the attachment of powerful neck muscles and the lower part of the face protrudes forward. The canine teeth are relatively large – those of males larger than those of females. The brain capacity is up to 24 in³ (400 cm³).

The skull of *Australopithecus afarensis*, the earliest known hominid species, is similar to that of the ape despite the fact that the rest of the skeleton of *A. afarensis* was dramatically altered for upright walking. The brain capacity is much the same – 23–27 in³ (380–450 cm³) – and the skull ridge and large canines are still present.

The modern human skull shows that in the 3 million years since *A. afarensis* lived, the brain has increased to about 82 in³ (1,350 cm³). This brain is encased in an almost spherical skull that has lost any trace of the crest or ridge. The lower face is less protuberant and the teeth more even in size.

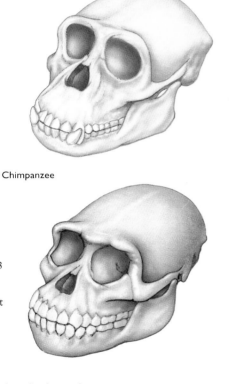

Chimpanzee

Australopithecus afarensis

The human skull is flat at the front with a large spherical braincase and is perched on top of a vertical neck. By contrast, the ape's skull has a relatively long snout region and is carried out in front of the neck vertebrae. The braincase has strong ridges for the attachment of jaw and neck muscles.

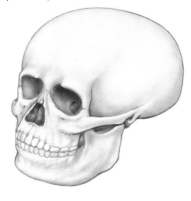

Modern *Homo sapiens*

THE RULES OF CHANGE

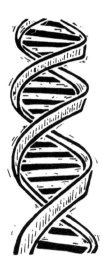

Life and evolution are intertwined. Since the beginnings of life on Earth they have been the most intimate of symbiotic partners and they continue to coexist in our world today. Life and evolution need each other.

By means of evolution, living systems change through time, the alterations happening in response to environmental change. In its widest sense of change driven by natural selection, evolution can operate in non-living systems – in interacting sets of chemical reactions for instance. But it is at its most supreme when it operates on the complex self-copying systems of living things.

Over a period of more than 3 billion years of accumulating change this intermeshing of evolution and the life of our planet has produced an extraordinary variety of life forms – organisms ranging from the tiniest bacteria to the mightiest dinosaurs and whales. That epic of change has been charted through fossil evidence and is manifest in the diversity of animals and plants that live now.

In order to understand the real processes which underpin those billions of years of evolution, some profound issues must be confronted. Life itself must be defined to see why it is irrevocably joined with evolution. The emergence of life from the non-life that preceded it must be charted and this demands an explanation of life in basic chemical terms.

Indeed, the modern analysis of evolution is now framed by our molecular understanding of its workings. Inheritance, sexual reproduction, natural selection and adaptation are all concepts that can now be defined in terms of the molecular biology of DNA itself.

Living things are self-regulating – they do not have to be externally programmed. A hypothetical equivalent of such a system would be a self-organizing automobile factory that obtained its own raw materials, designed its own blueprints, built the automobiles, marketed them, assessed market responses and then – most crucially – modified the designs to achieve improved sales in an ever-changing marketplace. For one of the brand marks of living things is their inherent capacity for making changes in themselves appropriate to their surroundings and then passing on these beneficial alterations.

These changes are passed on from parent to offspring as genes. Genes composed of DNA are the codes that an organism uses to regulate and maintain itself. The gene collection of every individual plant or animal organism is known as its genotype.

The sequence of molecular building blocks contained in the DNA is a code for – and can bring into existence – a corresponding sequence of amino acids. From these amino acids a specific protein is made. The sequence of amino acids determines a protein's characteristics and the way it functions within a living organism.

Modern molecular biology gives today's scientists the power to dissect the genetic and metabolic workings of living creatures at the most basic level of both atoms and molecules. An analysis of the molecular language of life – the language spoken by DNA, RNA and proteins – reveals an evolutionary drama whose stage directions are written in the genetic code.

Mutations in this code – changes in the DNA which lead to new proteins – provide the raw material for this drama and, from generation to generation through geological time, for the incredible diversity of life on Earth.

WHAT IS LIFE?

Trees, big cats, sparrows, even tiny ants and worms are obviously living organisms. But life on this planet is much more diverse than this. Fungal spores, green algae, microscopic organisms teeming in a droplet of water, single-celled protozoa, bacteria and viruses – all these are just as "alive" as more visible animals and plants. Even though they cannot be seen, the presence of these minute organisms is everywhere. A plant wilts, yellows and dies due to infection with invisible virus particles, bacteria causes milk to sour, common cold viruses make our noses run and eyes stream.

Despite their disparate forms, all living things share certain characteristics. These bind together all life forms from bacteria to baleen whales.

First and foremost, living things must reproduce. They make near-perfect copies of themselves in two main ways. The first is sexual reproduction, a system which at first glance seems unnecessarily complex. It requires that each type of organism, each species, exists as two separate forms: one with organs for producing single-celled eggs and another with an equivalent system for producing single-celled sperm. One egg has to fuse with one sperm to produce a compound cell – the fertilized egg or zygote – which can grow into a new independent organism with all the features of the parental species. A few organisms, such as slugs and snails, simplify this pattern by being hermaphrodite – each individual can make both eggs and sperm. Such organisms are, however, comparatively rare.

Plants and fungi employ the same cellular machinery but have different specialized cells. For all the organisms that use complex sexual reproduction the underlying rationale is the genetic variation produced (pp. 70–73).

The other method of reproduction is asexual. A single-celled organism splits into two genetically identical daughter cells, each of them a new organism of the original species. Many algae and protozoans such as amoebae, flagellates and ciliates multiply in this way. Some multicellular organisms, such as many plants, seaweeds, corals and worms, also reproduce asexually. Part of the organism splits off and grows into a new individual, genetically identical to the parent organism.

In order to survive and be able to reproduce, an organism must be maintained by four linked processes of life which constitute its basic machinery: growth, nutrition, respiration and excretion. For an organism to grow, new living material in the form of new cells is needed. New material cannot be conjured from nothing; it can only be produced by an organism acquiring raw materials from outside its body and transforming these into its own substance. This is the essence of nutrition.

This American diver is just one of thousands of species of bird that have evolved to colonize land, air and water habitats on Earth.

Mammals, such as the mighty leopard, have dominated life on land since the disappearance of the dinosaurs at the end of the Cretaceous, 65 million years ago.

Gymnosperms – plants such as conifers and spruce which bear their seeds in cones – dominated Earth before the evolution of flowering plants.

All living things need to take in nourishment and there are two main ways of doing so. Plants and photosynthetic and chemosynthetic bacteria take inorganic molecules, including carbon dioxide, from the environment and change them into organic molecules, the substance of the plant or bacterium.

Animals, fungi and many forms of bacteria require more complex food, however. They consume the live or dead bodies of other organisms or parts of the bodies that have been broken down. These organic foods are then converted into the particular organic molecules of the consuming creature. For example, a cow eats grass. The proteins in that grass are broken down into amino acids and transformed into the proteins that build the cow's own body. That protein may in turn be eaten by another creature.

To operate the process of nutrition and other parts of an organism's metabolic machinery, an energy source is needed. For all organisms, this source is a group of so-called energy-rich compounds, the most common of which is adenosine triphosphate or ATP. By the process of respiration all organisms construct ATP by adding an extra phosphate group to adenosine diphosphate (ADP).

Plant respiration is partly powered by energy from the sun, but most other organisms make ATP by gaining energy from the breakdown of substances such as sugars and fats into simpler molecules. In a yeast cell this is achieved anaerobically – without oxygen. Most other organisms, including humans, need oxygen to maintain aerobic respiration, a particularly efficient mode of molecule breakdown. In the cells of most animals, for instance, glucose is oxidized with oxygen to generate ATP, with carbon dioxide and water as by-products.

The metabolic activities of growth, nutrition and respiration produce many substances which can be recycled for additional growth or respiration. But they also produce wastes which cannot be processed further. In animals, such waste products as urea, uric acid and

Flowering plants or angiosperms, such as these California poppies, are the largest group of land plants. Angiosperms form the basis of most terrestrial ecosystems.

carbon dioxide are eliminated from the body by processes of excretion. Carbon dioxide is removed in exhaled air while urea, dissolved in water, is excreted in the form of urine.

For a living thing to get the nutrients it needs – and sunlight, oxygen and a mate when necessary – it must be able to respond actively to its environment. A plant needs to grow into the light to photosynthesize, a lion has to locate its gazelle prey. A mud-dwelling worm must seek the aerobic environment it needs and avoid oxygen-less black mud.

And all sexual reproducers must find appropriate partners so they can get their sex cells together. To achieve these ends an organism uses sensory systems to pick up information about the state of the environment around it, information about light, heat, sound, chemical constituents and physical contact. Using this, the organism can coordinate appropriate responses. For many creatures these include locomotion – the ability to move around by swimming, crawling, walking or flying.

These seven "signs of life" – reproduction, growth, nutrition, respiration, excretion, senses and locomotion – mean that most living things can be easily distinguished from non-living inanimate objects. But there are grey areas. It is questionable, for instance, whether viruses actually respire. And although most plants can bend toward the light they cannot move about like animals do. Both the AIDS virus and a redwood tree are, however, still living things.

Such considerations prompt a more fundamental definition of life, one that encapsulates what is now understood about the molecular basis of life forms. This definition also underscores the link between life and evolution.

Living entities have a unique capacity. Within their own structure they contain an information system which, if appropriately supplied with the necessary raw materials, can build a new copy of the whole entity, including the information system. This definition may even reach beyond the particular life we know on our planet and enable us to recognize exotic life forms which we assume have arisen on other planets.

The key interactions in this definition are encapsulated in the diagrams below, which are inevitably influenced by earthly life as we know it. The entities are bounded by a membrane-like container and their information store is coded in the form of a linear sequence of molecular sub-units. Both these characteristics are essential for life on Earth but may not be absolute necessities elsewhere. Organisms arising from non-earthly evolutionary sequences could have different sorts of interfaces with their environment; information stores could have a different conceptual basis.

The information store of an organism must be large in order to code for the complex molecular machinery of self-copying. Part of self-copying entails copying the information store and it is here that the umbilical link between life and evolution is manifest. In the real world it is impossible to perform a long sequence of absolutely specific

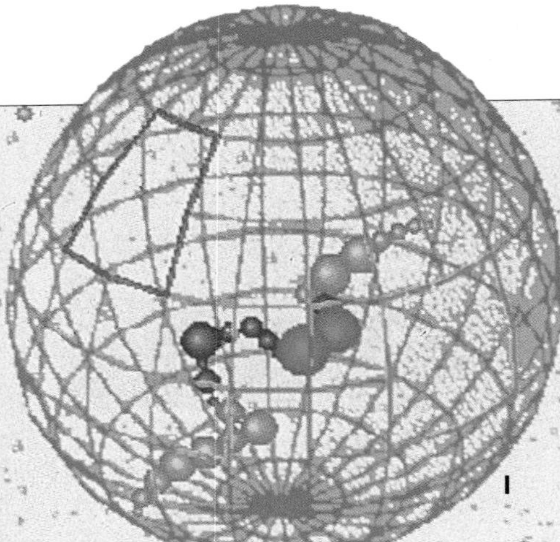

The basic enclosure and information store of the "organism".

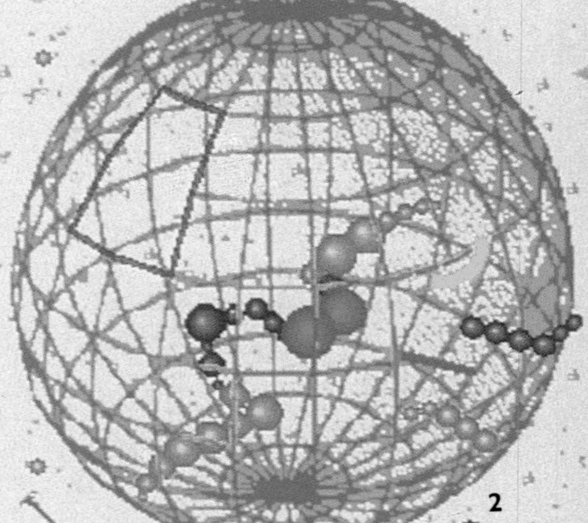

1

2

THE GROUND PLAN OF LIFE

All living organisms interact with the environment in much the same way. These schematic diagrams do not represent actual living things. Instead they illustrate the principles behind such interactions.

This basic organism consists of an enclosure of molecules – the green sphere – which serves to separate the living entity from its environmental surroundings.

There are two specialized zones in the enclosure. One (the blue opening) allows other molecules from outside to be incorporated into the structure of the organism – essentially the process of nutrition. The second (the red zone) is where new enclosure components can be

made into a copy of the organism – the process of reproduction.

Within the enclosure, the multicoloured chain represents the information store of the organism. This store has two major properties. First, it must be able to make an accurate copy of itself so that each new organism made during the process of reproduction can itself contain a version of the information store.

Second, the information store must be able to specify the construction of the molecules from which the organism is made up. It must also be able to direct the building of these components into a whole organism.

The incorporation of new molecules from the environment into the "organism".

instructions without errors occurring. The longer the sequence the more inevitable are mistakes.

Although the copying of life's information stores is extraordinarily accurate, things inevitably do go wrong and these mistakes are the raw material of evolution. Because of a chance copying error, an organism may be built that is slightly different from its parent. If this organism is in any way more efficient than its predecessors, the copies it makes of itself, with its subtly altered information store, will become more prevalent than those made by the parental type. Natural selection, with all its implications for life, is at work.

Corals may look plantlike but are actually colonies of animals. They live attached to the seabed, trapping food particles from the water.

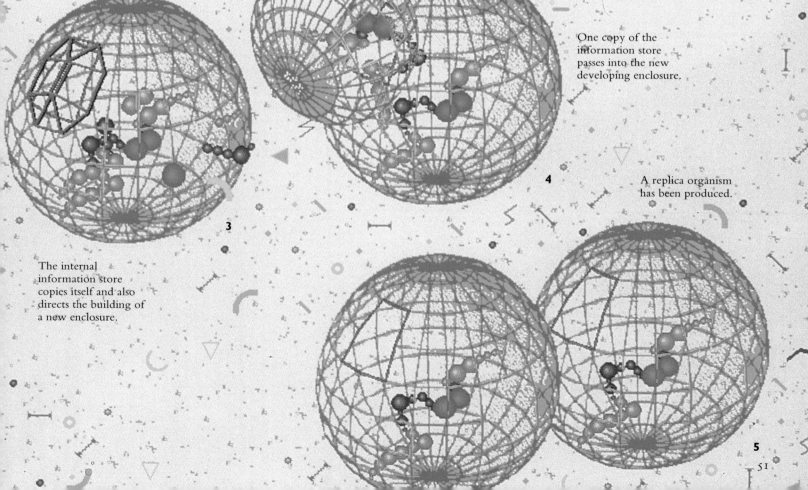

One copy of the information store passes into the new developing enclosure.

4

A replica organism has been produced.

3

The internal information store copies itself and also directs the building of a new enclosure.

5

ORIGINS OF LIFE

Four billion years ago there was no life on Earth. The world was being bombarded by huge meteors and there were no oceans. Life could not have existed in such conditions. But, by 3.5 million years ago, most experts believe that there were living cellular organisms on Earth. And those organisms must have been pre-dated by self-replicating systems that had not yet formed themselves into organized cells.

The self-replicating, information-carrying molecules of today – DNA and RNA – provide clues about the earliest self-copying molecules. In organisms now, the copying of RNA and DNA is achieved with the help of batteries of enzymes, each of which is a protein of some kind. But the system was not this complex at the outset. Of these two sorts of molecules, it was probably RNA (pp. 62–63) that came first. Once RNA or something like it had developed, subsequent evolution might have provided proteins and DNA.

Basic inorganic constituents, from which organic molecules could be built, were present in the early Earth's atmosphere. These included carbon dioxide, nitrogen and water vapour.

But where did RNA come from? With organic bases, sugars and phosphorus in its make-up, it seems too complex a molecule to have arisen in one chance step. In 1953, Stanley L. Miller of the University of Chicago cast light on the way complex polymers such as RNA may have been produced on the early Earth. He passed electrical sparks (simulating lightning or another energy source such as ultraviolet light) through the gases then thought to make up the early atmosphere of Earth – methane, ammonia and hydrogen. The ocean of the time was represented by water at the bottom of a sealed container.

In a few days, the water became discoloured with a reddish residue. This complex brew, formed from water vapour and the other gases by the energy of the electrical sparks, proved to contain amino acids – the building blocks from which proteins are formed.

Similar experiments carried out since have used a different gas mixture – water vapour, carbon dioxide and nitrogen – now believed to have made up the early atmosphere. The amino acids still form.

Energy sources, such as lightning strikes and ultraviolet radiation from the sun, changed the inorganic matter into simple organic molecules which dissolved in water.

Also produced are the organic bases, nucleotides, which are the key components of RNA and DNA. Since these bases, amino acids and other small organic molecules can be made without being produced by life forms, they could have accumulated in the gathering seas of early Earth. Today, such substances would be consumed as nutrients by microorganisms, but there were none in existence at that time.

The small organic molecules (monomers) had then to join up into complex chains (polymers) without the aid of catalytic enzymes to speed the process. Among many theories about this process is the idea that the reactions involved did not happen in free solution. The monomers could have been absorbed and aligned on the surfaces of solid material, such as clay particles, where side by side the combination of the monomers into short-chain precursors of RNA might have been made easier. This attachment and alignment could occur because of the crystalline ordering of electrostatic charges on the surfaces of mineral particles.

Organic bases joined together into chains of the earliest nucleic acids, probably RNA. This process may have been helped by happening on a solid surface such as clay particles (see below).

It seems likely that some of the randomly produced sequences of RNA-like polymers had a minimal catalytic copying ability. Thus, a type of natural selection could begin. Those sequences would have been better at taking monomers from the watery surroundings to make copies of themselves and would have become more numerous.

At some point in this process a final crucial step must have occurred. The self-replicating complexes enclosed themselves in bags, dividing themselves off from the surrounding water. Not least of the advantages this conferred was that the molecules that had been constructed would not float apart from one another. Today's cells are enclosed in baglike cell membranes made of fatty substances such as phospholipids. Similar molecules, with the ability to aggregate into thin sheets and form tiny hollow spherical bubbles, may have produced the membranes round the early self-replicators.

Once this had happened the world was inhabited with the very first cells. Recognizable life was on the Earth.

Volcanic eruptions spewed carbon dioxide, nitrogen and water vapour into the Earth's atmosphere. These were some of the inorganic subtances from which the earliest organic molecules were made.

Early nucleic acid sequences developed the capacity to copy themselves and eventually to control the synthesis of proteins.

RNA, protein and double-stranded DNA became incorporated into minute enclosures made of fatty molecules. The first cells had been formed.

The enclosure – a cell membrane – allowed interactions between nucleic acids and proteins to take place in a protected environment, relatively unaffected by external changes.

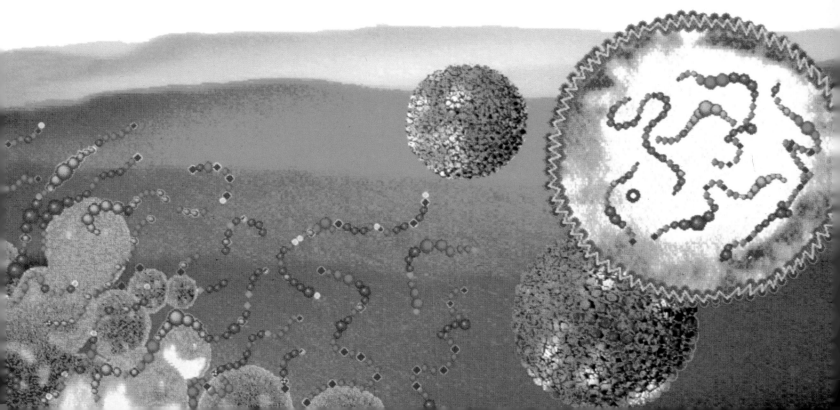

CONSTRAINTS OF LIFE

The existence of all living things on Earth is underpinned by the chemistry of carbon. But is this the only possible chemical basis on which life can be built, or are there other, "exotic" models that might exist elsewhere in the universe?

Most scientists are convinced that there is a universal chemistry. The atoms, atomic nuclei, ions and isotopes that can be examined directly on Earth, and indirectly in stars and other deep-space structures, come from a single atomic "tool-kit". Matter is matter everywhere: there is a universe-wide conservatism in the ways in which subatomic particles combine to form atoms. A similar conservatism dictates the ways in which these atoms can link up to form molecules and react with one another. Thus the understanding of atoms and molecules on Earth should be a valid basis for considering their behaviour anywhere in the universe.

In the construction of life forms – entities which contain an information store that codes for the copying of the whole entity, including another information store (pp. 62–63) – there are some atomic and molecular constraints. In particular, large precise information stores can be built only from large molecules with a capacity to code for other molecular structures. In Earth-dwelling organisms, the information in DNA molecules codes for the structure of smaller RNA molecules; their information content in turn codes for the structure of proteins, which make up the working machinery of an organism.

The coding capacity needed to build a large diversity of structural components can be achieved most economically by a sequence that uses a relatively small number of standardized sub-units. Here, simplicity is best. Any other molecular strategy would entail an impossible variety of coding sub-units and a consequent increase in the machinery required to recognize these components. With a sequence-centred code, it is possible to specify the structure of an almost endless number of protein molecules using only four DNA bases – the nitrogen-containing nucleotides which are the key components of DNA (pp. 60–61).

In living things the molecular information store appears to be constrained to consist of one or a few large molecules, made up of a number of sub-units. In these polymer molecules the various sequences of the four sub-units represent codes for a diversity of end-products.

Most life scientists believe that this apparent constraint limits the types of molecule that can attain these design specifications. In known chemistry, only carbon-based molecules seem to fit the bill. They have an almost limitless capacity to form huge polymers, and their combining abilities allow them to construct the necessary diversity of sub-units. Carbon also has the additional diversity-generating capacity of being able to link together with other elements to form complex molecules. The skeleton of organic molecules is a framework of carbon-carbon linkages but carbon's ability to link with oxygen, hydrogen,

nitrogen, sulphur, phosphorus and other atoms immeasurably expands its molecule-building repertoire.

The bonds between the carbons are generated by electron orbitals. Those of carbon are directed to the four corners of an imaginary tetrahedron with a carbon atom at its centre. This intrinsically three-dimensional arrangement means that organic molecules can expand and branch freely in all directions.

The possibility that other four-bonding (four-valent) atoms, such as silicon, might provide as much flexibility as carbon has been the subject of much research. As the diagrams here show, silicon can make some simple organic-like molecules, but in ordinary circumstances – in water and at normal temperatures and pressures – cannot achieve the more complex configurations of the carbon family.

Life on Earth is apparently limited to the use of carbon-based chemistry for building the machinery of a life form and its information store. Molecules of the general form of DNA, RNA and proteins seem essential for life. But in a life system that originated independently from our own, the sub-units of the information store and working molecules could be quite different.

The carbon-centred system of life on Earth also constrains the physical conditions in which the metabolic processes of life can happen. If molecules are to be broken down and reconstructed in a controlled way, if information codes are to be translated into working molecules and if information stores are to persist unaltered over a long period, there are two basic requirements: many of the molecules must be in solution and in

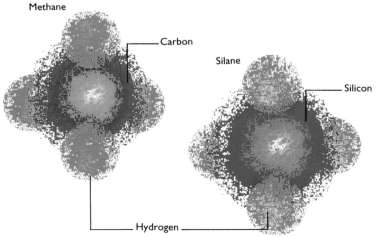

Ethane

Methane

Carbon

Silane

Silicon

Hydrogen

Methane (CH_4) is one of the simplest organic carbon molecules. It consists of a single carbon atom, surrounded by four hydrogen atoms. Silane (SiH_4) is the exact silicon analogue of methane – four atoms of hydrogen with a silicon atom at the centre.

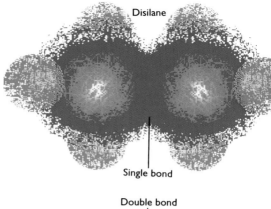

Ethane (C_2H_6) contains two carbon atoms linked by single bonds. Disilane is its silicon-based analogue. Ethylene (C_2H_4) differs from ethane in having a double bond between its two carbon atoms.

Disilane

Single bond

Double bond

Ethylene

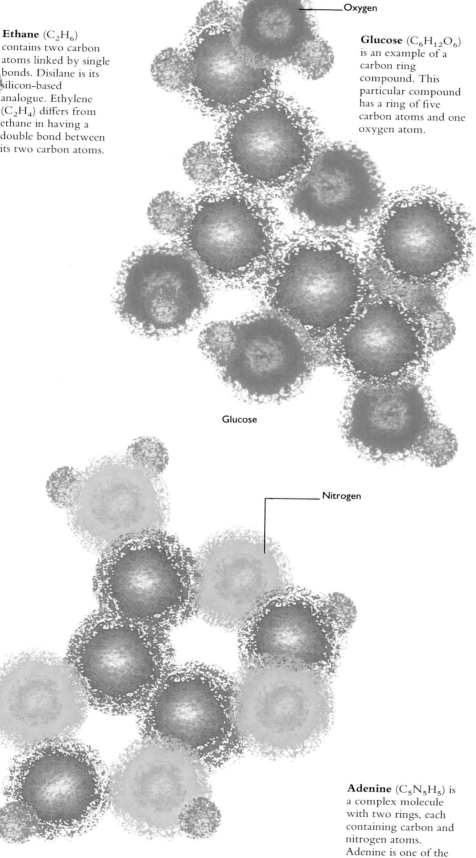

Glucose ($C_6H_{12}O_6$) is an example of a carbon ring compound. This particular compound has a ring of five carbon atoms and one oxygen atom.

Oxygen

Glucose

Nitrogen

Adenine ($C_5N_5H_5$) is a complex molecule with two rings, each containing carbon and nitrogen atoms. Adenine is one of the components of DNA.

Adenine

non-extreme conditions of temperature and pressure. Water is the only common general purpose solvent for dissolving organic molecules. To exist in the necessary liquid form, not as ice or steam, it must be at temperatures of between 32°F (0°C) and 212°F (100°C).

In the chemistry of this universe, therefore, it seems that life must be based on carbon and liquid water. This limits the places in the universe where life and evolution can occur. Cradles of life must be planets, moons or asteroids whose distance from a star and atmospheric conditions imply a temperature range of 32–212°F (0–100°C), at least in some regions. The physical structure of the planetary body must be one that possesses or generates water. And carbon must be present. If these conditions are maintained on a planet it is probable that, in time, life forms will develop. They will then slowly, but inexorably, diversify under the impact of natural selection.

Only time and a level of planetary exploration beyond our present expertise will reveal whether this is valid. At present Earth, unlike nearby Venus and Mars, fits the design requirements for a life-generating planet. But if life has begun on any other planet or moon in this or any other galaxy, what might 4 billion years of evolution achieve there?

THE SMALLEST UNITS

For some 3 billion years or more, throughout the Precambrian, evolution operated in a world of single-celled organisms. Each of these single cells was a living creature in its own right. And each species that adapted to a particular niche in that early world was made up of a population of such single cells – cells being, in most cases, the smallest units that can be termed living.

A cell is really a package of carbon-based molecules in a watery solution enclosed by a flexible boundary, or membrane. In all types of cell the membrane is only 10 nanometres thick – a pile of 10,000 cell membranes, one on top of the other, would equal the thickness of this page. This gossamer-thin membrane is composed of a double sheet of charged fatty molecules called phospholipids. Proteins attached to the membrane's inner and outer surfaces, and penetrating its thickness, form pores that allow chemicals to pass through.

The first successful cell types on Earth were much like today's bacteria and blue-green algae. Termed prokaryotes, these single-celled organisms – some grouped into lines or sheets – were, according to the microfossil record, the dominant life forms on this planet for much of the Precambrian.

Prokaryotic cells, such as bacteria, are so tiny that even under an optical microscope they are little more than dots. Under the electron microscope they are shown to possess a cell wall – a semi-rigid carapace outside the cell membrane – but no cell nucleus. A typical bacterial cell consists of a circle of double-stranded DNA, usually in the middle of a cell, embedded in cytoplasm that contains thousands of ribosomes where the all-important proteins are manufactured.

Present-day prokaryotes exhibit a wide variety of shapes, from tiny spheres and rods to long spiralled helices and chains of spheres. Despite the apparent simplicity of their structure, evolution has endowed them with extremely diverse metabolic capacities, making them critical participants in the flow of energy and molecules around the planet.

As the biochemical alchemists of life,

different prokaryotic species can photosynthesize like green plants, break down almost any form of organic molecule for use as a nutrient, or fix gaseous nitrogen from the air. They can generate methane, acquire their oxygen from sulphates and their hydrogen from hydrogen sulphide; they can survive and grow in the most extreme conditions of heat, cold and salinity.

Attempts by taxonomists to chart lines of evolution between the structurally similar, yet metabolically diverse microbes all but failed until they examined the DNA and RNA of the cells. The results of decoding the microbes' nuclear material have led to a proper classification of prokaryotes and, as

importantly, to the discovery of a new kingdom of life. Work on the so-called 16S fragment of prokaryotic ribosomal RNA revealed that bacteria could be separated into two evolutionary divisions – eubacteria and archaebacteria.

Eubacteria are the "normal" bacteria – if such varied organisms can be so designated. Archaebacteria, however, are particularly specialized to cope with the most extreme conditions that exist on Earth. Their metabolic biochemistries differ from those of eubacteria and their cell walls, enzymes and methods of photosynthesis are all unique.

Some archaebacteria are extreme anaerobes which means they can colonize oxygen-free environments, but

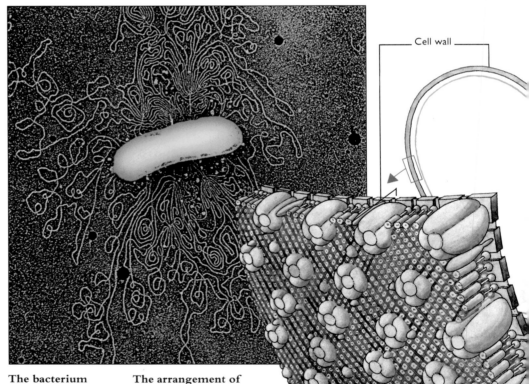

Cell wall

Matrix proteins

Lipid and phospholipid layer

The bacterium
E. coli, here seen at 23,000 times life size, lives in the human intestine. This bacterial cell has been treated with an enzyme to weaken its cell wall and then placed in water where the DNA, visible as fibrous threads, has been ejected.

The arrangement of cell wall, cell membrane and whiplike flagella on the outer surface of an *E. coli* cell is shown in the central image (*right*). The enlargements on either side reveal the internal organization of the cell wall and the cell membrane and flagellum base.

even the tiniest amount of oxygen is lethal to them. Others are sulphur-metabolizing forms that can grow and survive in extremely hot, acid environments. Yet others can live in superconcentrated salt solutions that would kill all other life forms. Clearly the archaebacteria are distinctive forms of life and well deserve their current placing in a separate kingdom.

Even more distinctive are the viruses: they are not cells and consequently do not readily fit into the taxonomic organization of life forms. Viruses are the smallest self-copying entities that contain DNA or RNA. It is as though they satisfy only the most basic, pared-down requirements for a molecular definition of life. A virus consists merely of a few genes which are housed inside a protein shell, or capsid.

Viruses cannot make copies of themselves without the help of other cells and can survive only by becoming parasites of individual host cells. As such, they are agents of disease in animals and plants. The tobacco mosaic virus parasitizes plant cells; HIV, the virus responsible for AIDS, parasitizes human cells; and viruses known as bacteriophages or phages (pp. 188–91) infect bacteria and blue-green algae.

Once the virus has attached itself to the host cell membrane, its genes pass inside. Here, they both duplicate themselves and generate new capsid proteins. These self-assemble into new viruses which, on leaving, cause the host cell to disintegrate and die. One by one, cells of a particular tissue are killed, with pathological consequences for the host organism. The herpes virus, for example, brings the discomfort of a cold sore, and HIV infection of the immune system can result in AIDS.

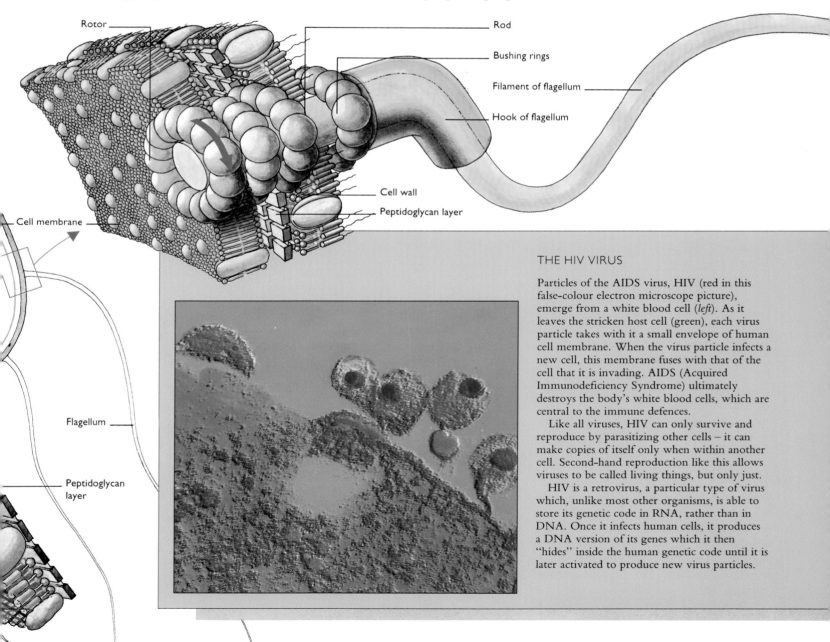

Rotor — Rod

Bushing rings

Filament of flagellum

Hook of flagellum

Cell wall

Peptidoglycan layer

Cell membrane

Flagellum

Peptidoglycan layer

THE HIV VIRUS

Particles of the AIDS virus, HIV (red in this false-colour electron microscope picture), emerge from a white blood cell (*left*). As it leaves the stricken host cell (green), each virus particle takes with it a small envelope of human cell membrane. When the virus particle infects a new cell, this membrane fuses with that of the cell that it is invading. AIDS (Acquired Immunodeficiency Syndrome) ultimately destroys the body's white blood cells, which are central to the immune defences.

Like all viruses, HIV can only survive and reproduce by parasitizing other cells – it can make copies of itself only when within another cell. Second-hand reproduction like this allows viruses to be called living things, but only just.

HIV is a retrovirus, a particular type of virus which, unlike most other organisms, is able to store its genetic code in RNA, rather than in DNA. Once it infects human cells, it produces a DNA version of its genes which it then "hides" inside the human genetic code until it is later activated to produce new virus particles.

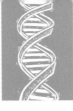

LIVING CELLS

The advent of more complex eukaryote cells, such as the first protozoans about 1.5 billion years ago, signalled a major turning point in evolution. Membranes that had formerly been confined to the boundary of simple prokaryotic cells developed inside the cell and enabled its contents to be organized into compartments. Thus cellular functions, such as DNA information processing, protein synthesis and ATP production, became increasingly sophisticated. This, in turn, meant that the cells of an organism had the potential to differentiate and so evolve highly specialized roles.

The inner complexity of the new eukaryotic cell type is thought to have evolved from a symbiosis, or sharing, between different types of prokaryotic cell. This symbiosis was particularly intimate since one cell actually came to live inside another (pp. 28–29).

The cytoplasm of a prokaryotic cell is a relatively simple suspension of ribosomes. A eukaryotic cell, by contrast, has a diversity of organelles – distinct, specialized zones of cytoplasm like subcellular organs, hence their name.

Eukaryotic cells, such as single-celled algae and protozoans, can live as independent organisms because their mix of organelles autonomously maintains all their activities. But in the evolution of multicellular fungi, plants and animals, the basic cellular unit has become differentiated into an almost infinite range of specialized cell types. This differentiation has, by means of evolution, resulted in specialization for different roles.

Thus, the organelle organization in a muscle cell enables it to contract; a gland cell is organized to manufacture and secrete a particular chemical, be it mucus or hormone. Both the muscle cell and the gland cell of one individual of one species will have exactly the same genetic constitution. Their differences in organization, however, will depend on the operation of two different sets of genes.

Eukaryotic cells could evolve highly specialized functions because membranes enabled them to create many organelle types. The eukaryotic cell membrane, though basically similar to that of a prokaryote cell, has new molecular components that allow more flexibility in terms of changes in shape and movement. In plants, the outer surface of the cell membrane is coated with a tough and rigid cellulose cell wall.

Mitochondria are double-membrane organelles; the enzymic machinery for the efficient aerobic production of ATP is carried on the inner membrane. Chloroplasts in plant cells are similarly multimembranous in organization, with the inner membranes bearing the enzymes and chlorophyll that are the basis of photosynthesis.

The systems for making, modifying and secreting an ever-increasing diversity of proteins are also centred on membranes. Double sheets of membranous vesicles, studded externally with ribosomes, make up the granular endoplasmic reticulum. Proteins made on the ribosomes pass into the endoplasmic vesicles and then to the vesicle stack called the Golgi apparatus where they pick up sugars. Other membranous organelles called lysosomes, which are filled with lytic digestive enzymes, break down unwanted cytoplasmic components or nutrients that have been taken up by the cell.

Somewhere in the centre of the cell's cytoplasm is a large, double-membranous enclosure – the nucleus. Within it are the chromosome pairs that carry all the cell's nuclear genetic information. Only eukaryotic cells compartmentalize their genes inside a nucleus in this way and have multiple chromosomal sets.

Important evolutionary steps were taken when the development of threadlike organelles known as microfilaments and microtubules made directional movement possible. These organelles transport materials across a cell, draw chromosome pairs apart at cell division and, as a flagellum on the male sex cell, power a sperm's movement toward a female egg.

ANIMAL AND PLANT CELLS

Cells come in an immense variety of shapes and types. In these examples of animal and plant cells, each cell and the organelles within it have a number of structures specialized for particular functions such as nerve impulses or muscle movements.

Cytoplasm

Granular endoplasmic reticulum

Mitochondrion

Dendrite

Nucleus

A sperm cell has an undulating flagellum powered by mitochondria. The acrosome, produced by a Golgi body, enables it to fuse with an egg cell.

This beta cell from the pancreas makes insulin on the granular endoplasmic reticulum. The hormone is then passed out of the cell in secretory inclusions.

Secretory inclusion with insulin

Inclusion releases insulin at cell surface

Mitochondrion

Granular endoplasmic reticulum

Nucleus

Golgi apparatus

Centriole

Acrosome

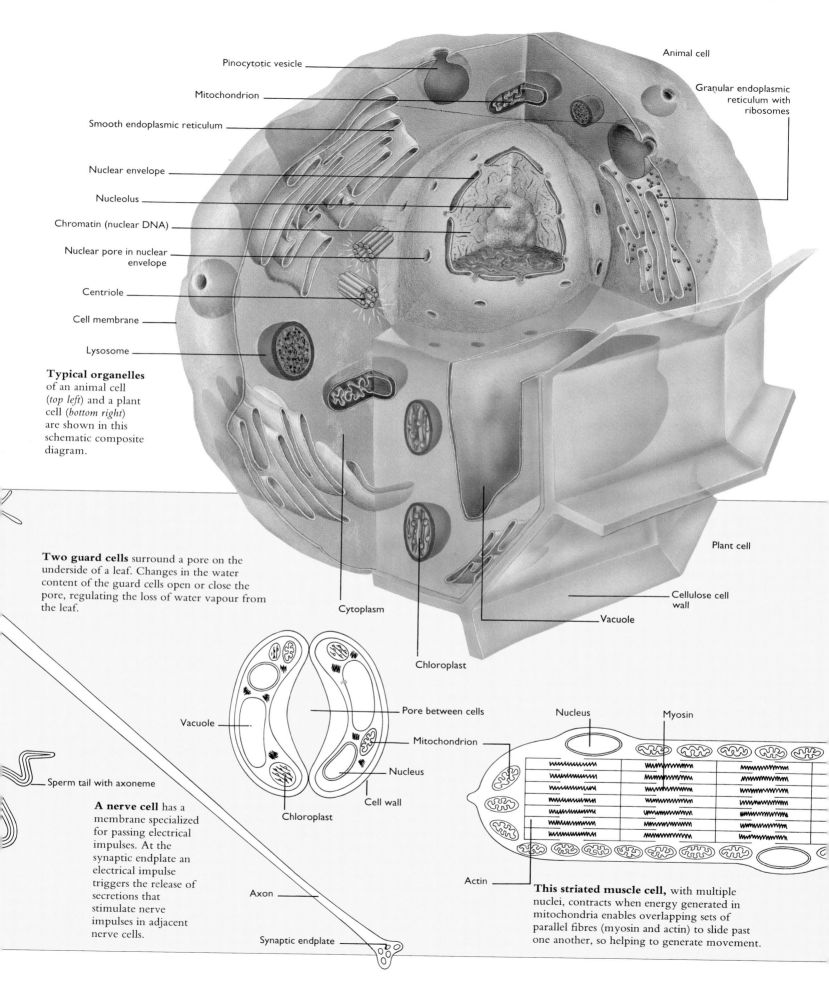

Pinocytotic vesicle

Mitochondrion

Smooth endoplasmic reticulum

Nuclear envelope

Nucleolus

Chromatin (nuclear DNA)

Nuclear pore in nuclear envelope

Centriole

Cell membrane

Lysosome

Animal cell

Granular endoplasmic reticulum with ribosomes

Typical organelles of an animal cell (*top left*) and a plant cell (*bottom right*) are shown in this schematic composite diagram.

Two guard cells surround a pore on the underside of a leaf. Changes in the water content of the guard cells open or close the pore, regulating the loss of water vapour from the leaf.

Cytoplasm

Chloroplast

Plant cell

Cellulose cell wall

Vacuole

Vacuole

Pore between cells

Mitochondrion

Nucleus

Cell wall

Chloroplast

Nucleus

Myosin

Sperm tail with axoneme

A nerve cell has a membrane specialized for passing electrical impulses. At the synaptic endplate an electrical impulse triggers the release of secretions that stimulate nerve impulses in adjacent nerve cells.

Axon

Synaptic endplate

Actin

This striated muscle cell, with multiple nuclei, contracts when energy generated in mitochondria enables overlapping sets of parallel fibres (myosin and actin) to slide past one another, so helping to generate movement.

DNA AND RNA

The publication of Darwin's *On The Origin of Species* in 1859 spurred many biologists to search for ways in which characteristics were passed down from parents to offspring. But unaware of the commotion caused by Darwin's book, the Moravian monk, Gregor Mendel, experimented on pea plants in his mon-

astery's garden and, in 1865, published his results on the inheritance of variation.

Mendel's conclusions went unnoticed for 35 years until workers in the field of breeding and inheritance discovered them in an obscure journal. It was then realized that Mendel had grasped the mechanical essence of genetics: that discrete portions of information were passed from one generation to the next. He had understood that these bits of inheritance, later to be called genes, pass unchanged from a parent to its offspring.

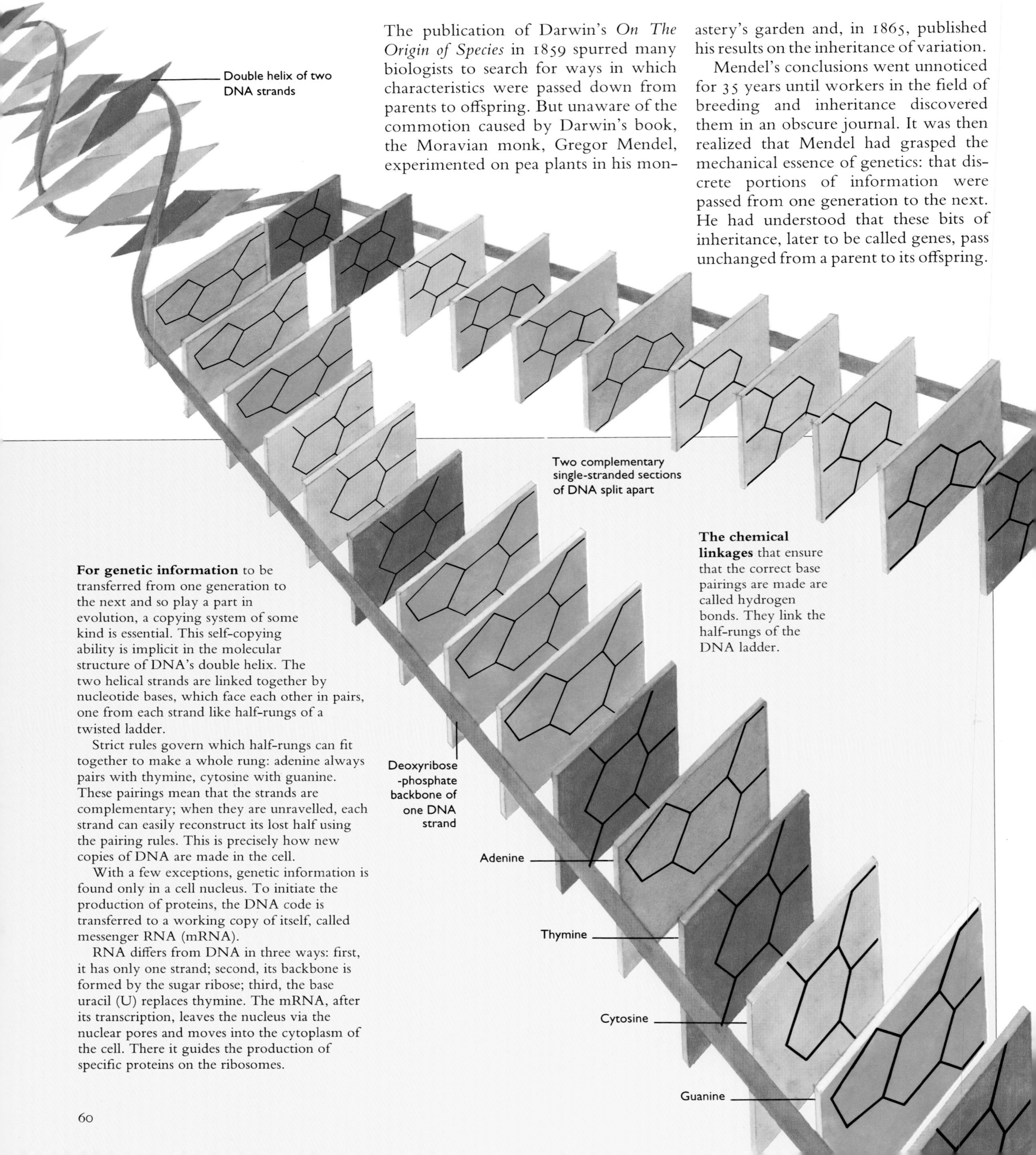

Double helix of two DNA strands

Two complementary single-stranded sections of DNA split apart

The chemical linkages that ensure that the correct base pairings are made are called hydrogen bonds. They link the half-rungs of the DNA ladder.

For genetic information to be transferred from one generation to the next and so play a part in evolution, a copying system of some kind is essential. This self-copying ability is implicit in the molecular structure of DNA's double helix. The two helical strands are linked together by nucleotide bases, which face each other in pairs, one from each strand like half-rungs of a twisted ladder.

Strict rules govern which half-rungs can fit together to make a whole rung: adenine always pairs with thymine, cytosine with guanine. These pairings mean that the strands are complementary; when they are unravelled, each strand can easily reconstruct its lost half using the pairing rules. This is precisely how new copies of DNA are made in the cell.

With a few exceptions, genetic information is found only in a cell nucleus. To initiate the production of proteins, the DNA code is transferred to a working copy of itself, called messenger RNA (mRNA).

RNA differs from DNA in three ways: first, it has only one strand; second, its backbone is formed by the sugar ribose; third, the base uracil (U) replaces thymine. The mRNA, after its transcription, leaves the nucleus via the nuclear pores and moves into the cytoplasm of the cell. There it guides the production of specific proteins on the ribosomes.

Deoxyribose -phosphate backbone of one DNA strand

Adenine

Thymine

Cytosine

Guanine

Before the discovery of Mendel's conclusions, German cytologist August Weismann knew that the genes were housed in the nucleus of every cell in a living organism. He was impressed by the privileged status and condition of these genes, particularly those inside the gamete cells, the eggs and sperm.

Gametes are the only cells that can carry genetic information from one generation to the next during sexual reproduction. When gametes – one of each type – fuse together, the first cell of a new individual is formed. Weismann noted that the genes of the "germ line" cells – those leading to the production of gametes – were the important ones. Both the cells and genes of an organism's body – its soma – are destined to die, but the genes of the gametes are passed on to the next generation. Weismann's crucial insight was to realize that what passes between generations is information, genetic information.

With the benefits of modern molecular biology, the discoveries of Mendel and Weismann can now be expressed in mechanistic molecular terms. Genes have a physical reality – they are made up of DNA molecules. The information transferred from one generation to the next is built into the structural arrangement of those DNA molecules.

DNA is deoxyribonucleic acid, an immensely long, double-stranded molecule with a unique organization. This is the famous double helix arrangement discovered by James Watson and Francis Crick in 1953. Each strand of the molecule is a helix, a long, corkscrew shape, and the two strands are wrapped together.

The connections between the two helices are at the heart of the way in which DNA can act as a carrier of information. To understand these connections, the architecture of the DNA must be unravelled. Each strand consists of a backbone of alternating sugar (deoxyribose) and phosphate sub-units. Attached to each sugar is a nitrogen-containing sub-unit called a nucleotide base. In all the countless trillions of DNA molecules that exist in the organisms of the world, there are only four different types of nucleotide base. They are called adenine (A), thymine (T), cytosine (C) and guanine (G). The A, T, C and G nucleotides can be thought of as genetic letters from which the language of genetic information is put together.

The language is constructed of words that are invariably three letters long: AAA, CGT, TCC and so on. This is the so-called triplet code. Ultimately each of a range of three-letter words is equivalent to, in production terms, a specific amino acid; amino acids are the building blocks of proteins; proteins make up the bodies of organisms and their biochemical enzymes and hormones. Via this chain of information, genetic codes can be transformed into living things.

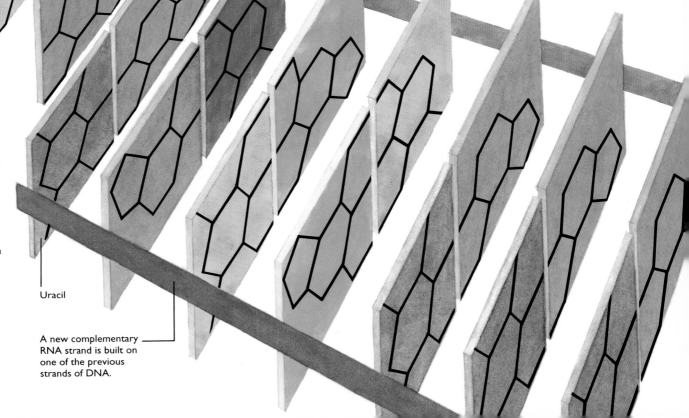

When the two complementary halves of a DNA helix have separated, one of the DNA strands is used as a template for building a corresponding molecule of mRNA. Complementary RNA bases bind to those on the single DNA strand via appropriate hydrogen bond links.

Uracil

A new complementary RNA strand is built on one of the previous strands of DNA.

The sum total of the genes in a cell nucleus represents a stupendous store of information. Known as the genome, this store will, for the most complex organisms, contain many millions of nucleotides arranged in a precise order. This arrangement, which makes a unique statement in genetic language, is identical in every nucleus of an individual. And it is this statement that makes each one of us a unique being.

The language of the genetic statement is found in the genetic code. From sequences of the four nucleotide bases of DNA – adenine (A), thymine (T), cytosine (C), guanine (G) – it is possible to make exactly 64 different three-letter code words, such as AAA, ATT, AGC and so on. The meanings of these 64 words are almost universally the same throughout the living world.

When put together, these words form codes which are transcribed into RNA and translated into proteins. One code, or gene, can be translated into one protein. This protein then performs a specific job in the cell or organism: for example, as a membrane protein, as a hormone or as one of the many enzymes that catalyse the biochemical reactions in the cell.

Three of the 64 possible words are used as "stop" signals to signify that the end of a protein code has been reached. The others code for the 20 different amino acids from which proteins are made. More than one word, in fact between one and six, can code for the same amino acid.

On average, a normal protein might be a precise sequence of several hundred amino acids. Given that three bases "equal" one amino acid, a sequence of one or two thousand bases on the DNA gives the code for one protein. Precision in base sequence becomes precision in amino acid sequence. Intriguingly, it seems that only a smallish fraction of nuclear DNA has the job of coding directly for proteins. The remaining genes probably help to regulate the workings of the DNA information store or else are operational during fetal development. They may conceivably be functionless in ordinary terms.

A typical chain of events links the DNA of a structural gene (one that codes for a specific protein) to eventual protein synthesis on the ribosomes. The gene sequence for the protein must first be "activated", in the sense that mRNA is transcribed by using the gene as a template (pp. 60–61).

In eukaryotes, the structural genes have a remarkable characteristic. The sequence of triplets coding for amino acids is interspersed by stretches, often of considerable length, which code for nothing. These "spacer" regions of DNA are called introns and the coding regions they separate are called exons. When an mRNA transcript of the whole structural gene is made, the sections corresponding to the introns are spliced out by enzymes before the mRNA passes into the cytoplasm to be translated into an amino acid sequence.

Once the mRNA copy of a structural gene has been made (minus the introns), it leaves the nucleus via exit points, or pores, in the nuclear envelope. The large diagram on the right shows, in a semi-schematic way, how translation of the mRNA takes place on the surface of ribosomes in the cytoplasm.

Ribosomes are tiny organelles, about 25 nanometres (a millionth of an inch) across, that are made of another type of RNA, ribosomal RNA, and also many enzymes. The ribosome is a translating machine which reads, in order, the triplet code sequences of the mRNA molecule, using each triplet to specify the appropriate amino acid in a growing protein chain.

The crucial partner in this process is a family of yet another type of RNA molecule – the transfer RNAs (tRNAs) – each of which can physically translate a codon, a code word of three letters, into the amino acid it stands for. The tRNA achieves this by having at one end the anti-codon, or triplet of bases, that pairs up with the codon. Attached to the other end is the amino acid for which the codon stands. Thus, with the tRNAs as coupling devices, a ribosome can translate an mRNA triplet sequence into an amino acid sequence.

INTRONS

A surprisingly large amount of an organism's DNA does not appear to be actively involved in coding for proteins. This proportion varies from species to species – for example, it is 9–27 percent in humans, 75 percent in nematode worms, 67 percent in fruit flies and 99 percent in lungfish. Some of this non-coding DNA is made up of introns. These sections of DNA seem to act as "padding" between exons – portions of DNA with a functional code. Introns are present only in the genomes of complex organisms. Bacteria, and other prokaryotes, do not possess them – almost all their genome codes for proteins.

The exons may, it is argued, be ancient "minigenes" that might once have coded for tiny functional stretches of amino acids. Modern proteins might therefore have resulted from the tacking together of groups of these exon-coded amino acids, and introns could simply have been part of a mechanism for patching these minigenes together into larger aggregates.

When mRNA is produced from a DNA gene, the RNA versions of the introns are created as a matter of routine. But before the mRNA can be used for protein production the introns have to be removed. They are "spliced out" and the exons linked together in the correct sequence.

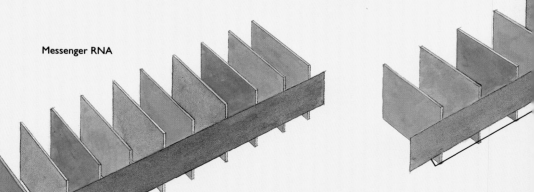

Messenger RNA

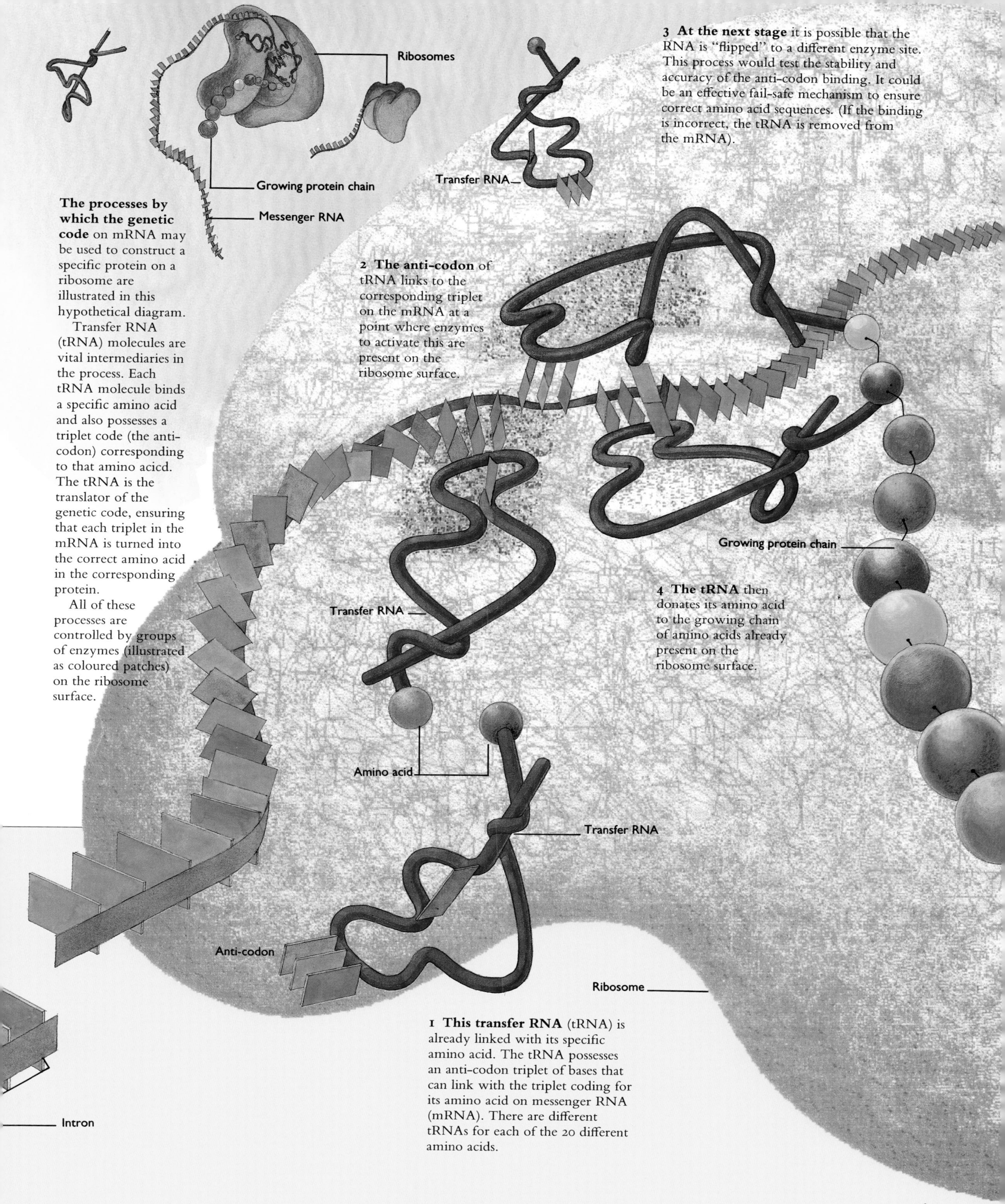

Ribosomes

Growing protein chain

Messenger RNA

The processes by which the genetic code on mRNA may be used to construct a specific protein on a ribosome are illustrated in this hypothetical diagram.

Transfer RNA (tRNA) molecules are vital intermediaries in the process. Each tRNA molecule binds a specific amino acid and also possesses a triplet code (the anti-codon) corresponding to that amino acicd. The tRNA is the translator of the genetic code, ensuring that each triplet in the mRNA is turned into the correct amino acid in the corresponding protein.

All of these processes are controlled by groups of enzymes (illustrated as coloured patches) on the ribosome surface.

3 At the next stage it is possible that the RNA is "flipped" to a different enzyme site. This process would test the stability and accuracy of the anti-codon binding. It could be an effective fail-safe mechanism to ensure correct amino acid sequences. (If the binding is incorrect, the tRNA is removed from the mRNA).

Transfer RNA

2 The anti-codon of tRNA links to the corresponding triplet on the mRNA at a point where enzymes to activate this are present on the ribosome surface.

Transfer RNA

Growing protein chain

4 The tRNA then donates its amino acid to the growing chain of amino acids already present on the ribosome surface.

Amino acid

Transfer RNA

Anti-codon

Ribosome

1 This transfer RNA (tRNA) is already linked with its specific amino acid. The tRNA possesses an anti-codon triplet of bases that can link with the triplet coding for its amino acid on messenger RNA (mRNA). There are different tRNAs for each of the 20 different amino acids.

Intron

63

GENETIC PACKAGING

The way the awesome quantity of genetic information is organized in the cell nucleus has profound implications for the evolutionary process. The DNA must be packaged effectively to make it not only accessible for mRNA construction (for subsequent protein synthesis) but also available for self-replication prior to cell division and sex cell production.

Though only a few thousandths of a millimetre across, a nucleus of a human cell contains an estimated 6.8 billion nucleotide bases, a length of DNA molecule which, if unravelled into a straight line, would stretch 10 ft (3 m). In a human this length is organized into 46 segments called chromosomes. Of these, 23 come from the sperm of an individual's father and 23 from the mother's egg that the sperm fertilized. On average, each human chromosome contains 130 million nucleotide bases.

The immense length of the DNA on the chromosomes has evolved to fit the tiny nucleus. A chromosome seems to consist of a DNA molecule associated with protein and RNA molecules that aid its packaging and function. Chromosomes are large enough to be seen under an ordinary optical microscope. In fact, they gained their name (from the Greek for "coloured body") because, due to their high DNA content, they can be easily stained during microscopic preparation. They change their appearance at different stages of a cell's activity. When the cell is not dividing and mRNA is being formed from parts of the genome, the chromosomes are largely tangled up, so they cannot be viewed separately.

During mitosis (see box), however, chromosomes copy themselves and double into two sets. They contract into thick, sausage-like cylinders that can easily be seen and counted, and it is now the chromosome complement of a species can be determined. The number and shape of this species-specific chromosome set is called the species karyotype. The chromosome number in a normal nucleus is called the diploid number (2N); the half set found in eggs and sperm is the haploid number (N).

These numbers vary greatly between different multicellular organisms. In humans, 2N = 46 and N = 23; in chimpanzees, 2N = 48. In the nematode roundworm, *Parascaris*, 2N = 2; in the hermit crab, *Eupagurus*, 2N = 254; and in some types of fern 2N = 500 or more. No simple relationship exists between an organism's chromosome number and its complexity.

Each chromosome in the set has a characteristic shape in a specific species. Only two basic shape types (plus intermediates) are possible and they depend on the position of the centromere, a kind of attachment point for the fibres that move chromosomes around during mitosis. Acrocentric chromosomes have the centromere at one end; in metacentric forms, the centromere is somewhere near the centre, making the chromosome into a V shape when it is moved around.

In some groups of organisms the chromosome number is constant between different species; for instance, 2N = 36 in most snakes and 2N = 26 in almost all dragonflies. In other groups, however, there may be extreme variations, with differences observable even between subspecies of one species. An unusual example is the muntjac, *Muntiacus muntjac*: the Chinese subspecies has a diploid number of 46 while that of the Assam subspecies is only 6.

Clusters of histone proteins

Nucleosome – about 180 base pairs wrapped round a cluster of histones

Coiled tube of linked nucleosomes

Double-stranded helical DNA

Deoxyribose-phosphate backbone of one DNA strand

Cytosine

Adenine

Thymine

Guanine

In chromosomes, the double helix of a two-stranded DNA molecule in the nucleus is packaged in a highly ordered way. Although each chromosome contains a single, immensely long DNA double helix, it is coiled and supercoiled and super-supercoiled in order to pack it tightly enough to form a chromosome.

Histones are clusters of proteins vital to the DNA packaging problem. The DNA segments are coiled around histone molecules into bundles called nucleosomes. The nucleosomes themselves are then coiled into yet more condensed structures – the chromosomes that can fit into a nucleus.

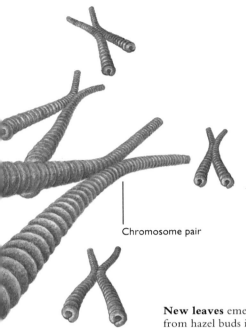

Chromosome pair

New leaves emerge from hazel buds in spring, each one the product of many thousands of mitotic cell divisions.

1 The nuclear envelope breaks down and the chromosomes start to condense.

Condensing chromosomes

Nuclear envelope breaks down

2 The chromosomes condense further and can be seen to be arranged in pairs.

Condensed chromosome pairs

3 The spindle forms and two identical sets of chromosomes begin to move toward opposite ends of the cell.

Spindle fibres

Metaphase plate of chromosomes

Centriole

5 In the two new daughter cells identical nuclei re-form. Each is surrounded by a new nuclear envelope.

Two daughter nuclei with nuclear envelopes re-forming

Separating chromosome sets

4 The two sets of chromosomes are now completely separated from one another. They have been pulled apart by fibres in the spindle.

Division furrow between two daughter cells

MITOSIS

Growth and development in many-celled organisms would be impossible without cell division. Over and over again individual cells must turn into two cells. Before a cell can divide, the genome – the total gene set – in its nucleus must copy itself and then split into two identical sets. Copying and splitting happen at different stages of the cell cycle. Copying takes place in the synthetic, or S, phase when each chromosome doubles up lengthways to create two identical halves, or chromatids.

Only after this can the splitting – the mitosis, or M, phase – take place. The spindle, a scaffolding of fibres that forms after the nuclear envelope breaks down, moves each chromatid of a pair to opposite ends of the cell. As a result, two new nuclei, identical to the originals, are reconstituted.

GENES AND ENVIRONMENT

When Gregor Mendel first explained the results of his breeding experiments with peas (pp. 60–61), he described the plants as containing two factors, each playing a potential role in controlling a specific trait, or characteristic. Modern knowledge about cellular and molecular biology has enabled Mendel's factors and their effects to be described in more concrete terms, particularly in the way they shape evolution and heredity.

Mendel arrived at his conclusions working only with what he could see of the pea plants – their nature and appearance, or phenotype. Today's scientists have become equally familiar with the genetic structure, or genotype, of an organism. They also know that an organism's phenotype is determined partly by its genes and partly by its environment and experience.

Mendel's factors are genes, specific sections of chromosomal DNA which represent the structure of a specific protein. The twofold nature of Mendel's factors is, in reality, a result of chromosomal arrangements. Each cell nucleus contains chromosomes in exactly matched pairs, matched because they contain equivalent sets of genes in the same sequence. In other words, a chromosome pair contains two genes for each characteristic. And, as modern geneticists have proved, a simple characteristic is determined by the production of a single protein.

One of Mendel's classic experiments in crossing pure-bred, tall and short pea plants reveals how the double gene pattern and its behaviour in sexual reproduction have profound consequences for heredity and evolution. When he crossed tall and short plants, all the offspring were tall; no short and no intermediate heights were produced. He then crossed the tall offspring with each other but they did not breed true. Their offspring were in the approximate proportions of three tall to one short.

These fundamental results imply that, in this example at least, inheritance of

Studies on the yarrow plant, *Achillea borealis,* in California reveal the separate yet overlapping influences of environmental factors and genetic make-up on a plant's phenotype or appearance.

In experiment 1, genetically identical clones of *Achillea* were grown not only at Mather, their site of origin, but also at sea level (Stanford) and at 10,000 ft (3,000 m) in the Sierra Nevada mountains at Timberline. Although all these plants were genetically identical, they grew to different sizes and shapes at each of the three locations. These variations are apparently caused by the different environmental influences – variations in temperature, soil and so on – at each of the three sites.

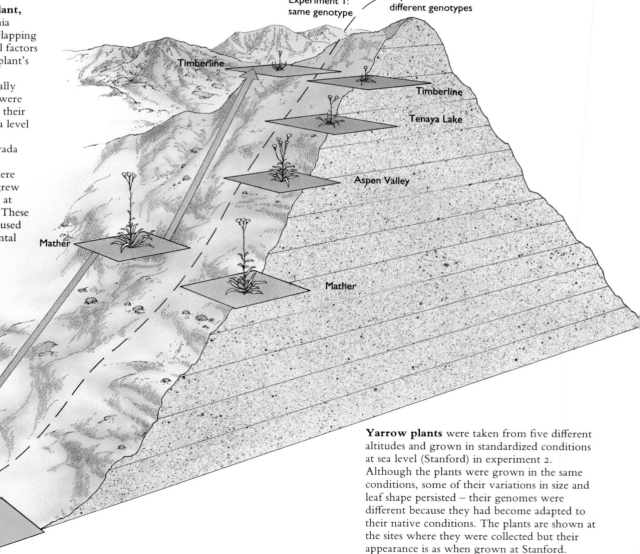

Yarrow plants were taken from five different altitudes and grown in standardized conditions at sea level (Stanford) in experiment 2. Although the plants were grown in the same conditions, some of their variations in size and leaf shape persisted – their genomes were different because they had become adapted to their native conditions. The plants are shown at the sites where they were collected but their appearance is as when grown at Stanford.

height must be passed on in defined units: since there were no plants of intermediate height, no blending of the characteristic occurred. The results also show the two-gene aspect of inheritance in operation. If the first tall plants have a gene for tallness (T) on each chromosome, and the short plants have a gene for shortness (t) on each chromosome, Mendel's results can be explained – as long as T and t interact in a particular way.

When a pure tall plant (TT) is crossed with a pure short plant (tt), each of the offspring receives a T gene from the tall plant and a t gene from the short plant. Their constitution will be Tt. But if, when they exist together, the effects of the T gene outweigh or take precedence over those of the t gene, it follows that all the Tt individuals will be tall. The T gene is said, in modern genetic parlance, to be dominant and the t gene recessive.

When two Tt plants are crossed with each other, they are just as likely to pass on their T gene as their t gene. The offspring will have equal chances of having a genotype of either TT, Tt, tT or tt. Because T is dominant, the first three possibilities will result in plants with a tall phenotype; only the fourth will result in plants with a short phenotype.

The colour of flowers is determined by their genes. Some of these are "structural" genes which are directly concerned with producing colour – genes for the intensity of red or yellow, for example. Others are control genes that switch the activities of other genes on or off.

Antirrhinums, as shown here, are often used in experiments on the genetics of flowers.

The results reveal an important relationship between genotype and phenotype. TT and Tt plants have different genotypes, but they have the same phenotype: they will both, given water, soil and sunlight, grow to be tall. Their genetic differences will only become apparent in the phenotypes of subsequent generations.

While one gene directs the making of one protein, many key characteristics result from the combined actions of several genes. These characteristics, such as height in humans, are said to be polygenically inherited. In these cases, more subtle outcomes may arise, including phenotypes that are intermediate to those of the parents.

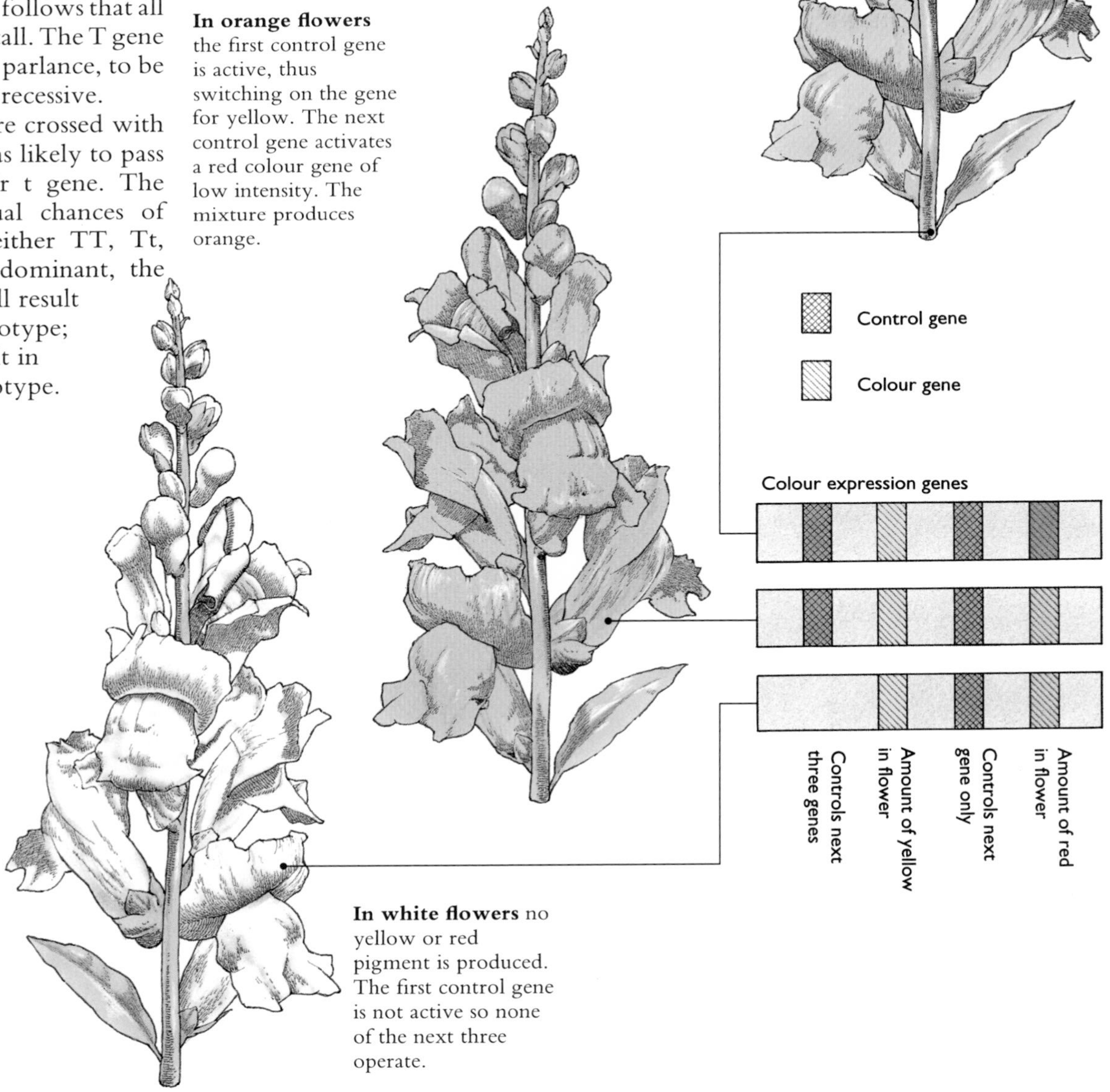

In orange flowers the first control gene is active, thus switching on the gene for yellow. The next control gene activates a red colour gene of low intensity. The mixture produces orange.

When a gene for yellow and a gene for high intensity red are both activated, red flowers result.

Control gene

Colour gene

Colour expression genes

Controls next three genes

Amount of yellow in flower

Controls next gene only

Amount of red in flower

In white flowers no yellow or red pigment is produced. The first control gene is not active so none of the next three operate.

SHUFFLING THE GENETIC PACK

As every gardener knows, plants such as strawberries can send out runners which grow into replicas of themselves. These new plants are genetically identical to their parents. The only genetic process involved is cell copying, or mitosis, which produces clones of the original.

Such clone-perfect similarity between parents and offspring certainly does not characterize sexual reproduction. The children of any two human parents, for example, are recognizably human and related to their parents – but they are also genetically unique.

This originality is the result of the genetic mechanisms which, by ensuring the diversity of offspring, are part of the "raw material" of evolution. The reproductive goal of a sexual encounter in any species is fertilization and the combination of one set of genes from one parent with the analogous set from the other. This takes place when the male sperm fertilizes the female egg.

In the human body every ordinary cell carries the same pattern of genes on 23 pairs of chromosomes. One chromosome of each pair came from the mother, one from the father. If the sex cells were made by mitosis, fertilization would result in the doubling of the chromosome number in each generation. Instead, sex cells are made by a specialized form of cell division called meiosis. In this process, the sex cells produced do not contain a full genetic complement, but a perfectly halved set – 23 chromosomes. When, at fertilization, one halved set in a sperm pairs up with another halved set in an egg, the chromosome number is restored to 46.

But this alone does not explain why every one of the 5 billion humans on this planet is genetically different (with the exception of identical twins). The chromosomes in the sex cells are not copies of the parent chromosomes. If they were, the opportunities for diversity in offspring would be limited. During the formation of a sex cell, the corresponding clusters of genes on each chromosome in a pair are shuffled together to produce a new combination of gene sets. The novel gene sets of all 23 chromosomes form the nuclear material of the sex cell. The two gene sets that a child inherits from its parents represent a unique mix of characteristics.

Identical twins, however, have exactly the same genes because they come from the same fertilized egg. For some reason, as yet unknown, the embryo divides at an early stage of development and grows into two babies.

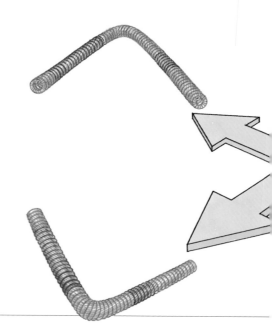

TWINS

The reasons why some twins are identical, some non-identical, can be explained by differences at the earliest stages of development.

Identical twins result from a single fertilized egg. But early in its development the resulting embryo splits into two. This makes two genetically identical embryos that will grow into identical twins, each with exactly the same genetic make-up.

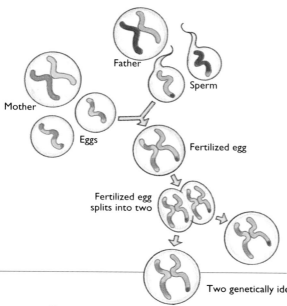

Mother
Father
Sperm
Eggs
Fertilized egg
Fertilized egg splits into two
Two genetically identical embryos

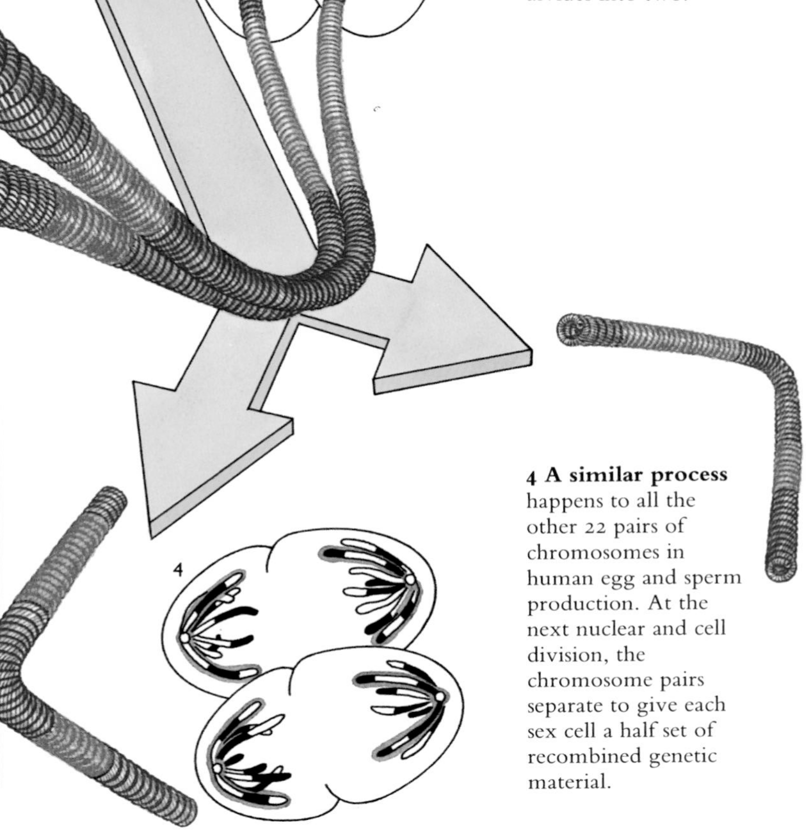

Sex cells – eggs and sperm – are made in ovaries and testes respectively by a genetic process known as meiosis. Each chromosome pair in a specialized cell duplicates itself and then shuffles gene sets at certain crossover points, producing four different half sets of chromosomes. Each egg and sperm has one of these half sets of genetic material – 23 single chromosomes in a human. At fertilization, when egg and sperm fuse, the number is restored to a full set of 46 chromosomes.

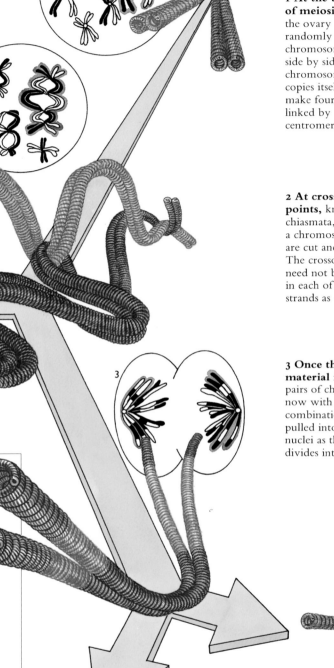

1 At the beginning of meiosis in a cell of the ovary or testis, the randomly distributed chromosomes pair up, side by side. Each chromosome pair then copies itself exactly to make four strands linked by a centromere.

2 At crossover points, known as chiasmata, sections of a chromosome pair are cut and swapped. The crossover points need not be identical in each of the four strands as shown here.

3 Once the genetic material is shuffled, pairs of chromosomes, now with new gene combinations, are pulled into two new nuclei as the cell divides into two.

4 A similar process happens to all the other 22 pairs of chromosomes in human egg and sperm production. At the next nuclear and cell division, the chromosome pairs separate to give each sex cell a half set of recombined genetic material.

Non-identical twins come from two separate eggs that by chance are fertilized at the same time.

Each embryo of non-identical twins is genetically different. The babies may be born at the same time but are no more similar genetically than any other normal siblings.

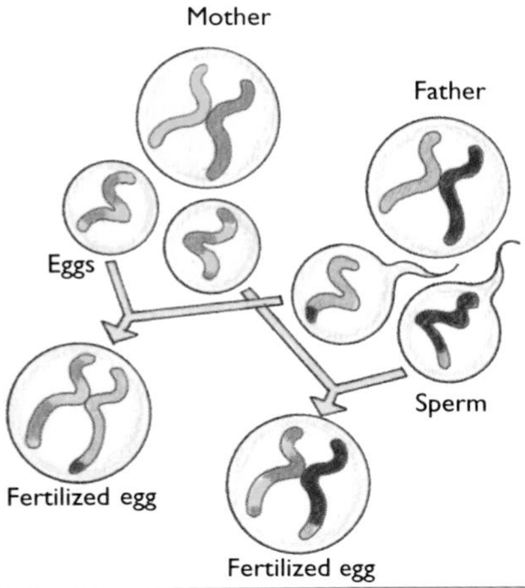

Mother

Father

Eggs

Sperm

Fertilized egg

Fertilized egg

WHO NEEDS SEX?

At the moment of fertilization, new evolutionary potential is released. When a sperm fertilizes an egg, or a pollen grain a plant's ovule, genes from both parents are combined in sexual reproduction and a new, unique individual is born. Yet the very commonness of sex and sexual reproduction among multicellular organisms is liable to inhibit what should be two key questions: why do they occur at all and how can they be selected for in evolutionary terms?

In the life cycles of most eukaryotes there is a shuttling, or alternation, between the diploid (full) and the haploid (half) chromosomal states. In most complex multicellular animals, such as humans, the haploid state is restricted to the gametes, eggs and sperm. The cells of all other developmental stages, whether they be larval, fetal or adult, have the full diploid composition.

In higher plants, such as the more advanced flowering plants, the haploid-diploid pattern is similar. Only the male pollen tube and the female ovule are haploid, the rest of the plant is diploid. Non-flowering plants, however, can alternate, in the same life cycle, between two generations – a haploid plant (the gametophyte) and a diploid plant (the sporophyte) (pp. 38–39). In mosses, for example, the main plant is a gametophyte. In some ferns the familiar leafy plant is a sporophyte; the haploid plant that alternates with it is a tiny structure that looks like a liverwort.

Whatever the life-cycle pattern employed for sexual reproduction it is normal for male and female gametes to be different. (Only in a range of single-celled protozoans and algae are the gametes identical.) Male gametes, such as sperm, are small, motile and made in large numbers. Female gametes, such as eggs, are larger and do not move; only a few are produced and these are often packaged with nutrient stores to power the early growth and development of embryos after fertilization.

In life cycles where the haploid gametes are not identical, it is the fertilized egg (now called a zygote and restored to a diploid condition) that produces a new individual. There is an inherent asymmetry in this situation: given the correct processing (normally, fertilization by a sperm) an egg can make a new individual, but a sperm cannot. Yet if eggs always had to be fertilized to develop further, and if sexual reproduction was the only way of making new individuals, sex would be obligatory for all organisms.

In fact, new individuals can be produced without any sexual process. Even when the basic gamete-producing machinery is in operation, fertilization is not always essential. In a kind of asexual reproduction known as parthenogenesis ("virgin birth"), practised by such organisms as aphids, a new individual will develop from an unfertilized egg..

Young produced by virgin birth (parthenogenesis) surround these adult female aphids.

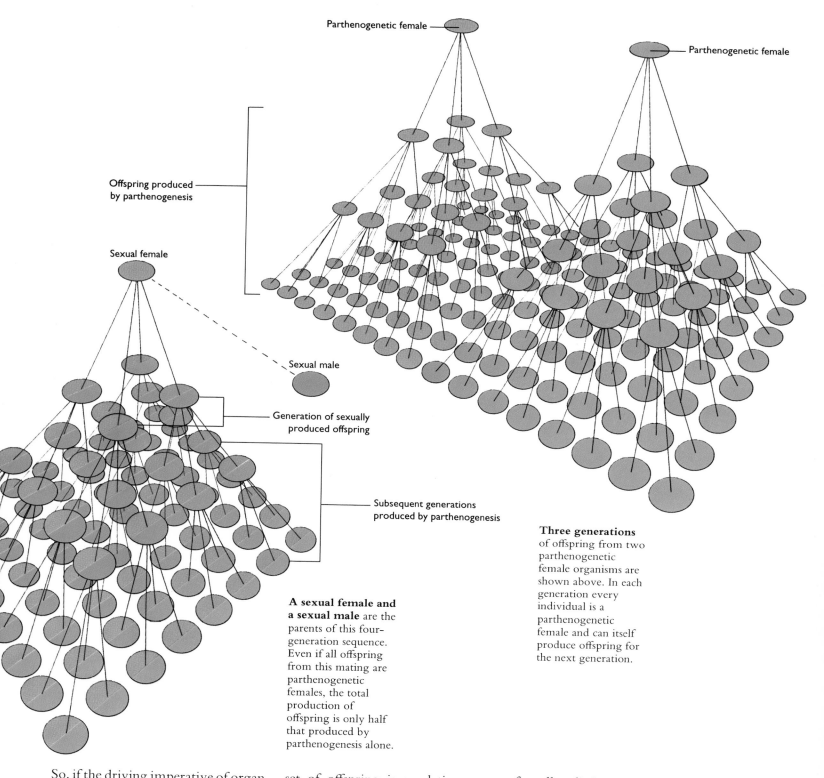

Parthenogenetic female

Parthenogenetic female

Offspring produced
by parthenogenesis

Sexual female

Sexual male

Generation of sexually
produced offspring

Subsequent generations
produced by parthenogenesis

A sexual female and a sexual male are the parents of this four-generation sequence. Even if all offspring from this mating are parthenogenetic females, the total production of offspring is only half that produced by parthenogenesis alone.

Three generations of offspring from two parthenogenetic female organisms are shown above. In each generation every individual is a parthenogenetic female and can itself produce offspring for the next generation.

So, if the driving imperative of organisms is to reproduce themselves and their genes as many times as possible in their lifetime, why is sex apparently the favoured way of doing it? On the face of it, asexual reproductive strategies such as parthenogenesis should always be favoured in an evolutionary, head-on competition with sexual reproduction.

The simply stated reason for this apparent puzzle is that, as males do not produce new individuals, any male in a set of offspring is a relative waste of reproductive output.

In parthenogenesis, a female produces female offspring that can, in turn, produce similar females in subsequent generations. Such a breeding system will obviously produce offspring, and thus gene copies, at about twice the rate of a sexual system in which males constitute half the offspring of each generation (see illustration). And efficient forms of asexual reproduction, such as budding and cell splitting, might also be able to produce offspring and gene copies faster than sex.

But, although used successfully by aphids and other creatures, such breeding systems as parthenogenesis have failed to pass the test of natural selection, otherwise they would have taken over the world long ago. The reason they have not done so must be that there are strong selective forces preventing this from happening.

One way of tracking these pro-sex selective forces is to look at the various types of reproduction that are not straightforwardly sexual and to identify what restricts their spread through populations.

In addition to parthenogenesis, there are other methods of asexual reproduction which completely exclude the use of gametes. New individuals grow either from the mitotic multiplication of a single normal body cell or a group of normal body cells. Strawberries which send out runners from the main plant to make new plants, the bulbs and corms of plants, and the buds of corals are all examples of such asexual reproduction. Unless a genetic mutation occurs in the process, the offspring are genetically identical to their parents. In other words, these organisms are making clones of themselves.

Although parthenogenesis has many complicated variants, it is patchily distributed among present-day eukaryote groups. Where parthenogenetic species do exist they usually have closely related species which are either sexual in their habits or use parthenogenesis alongside normal sexual reproduction in the same life cycle.

Mammals and gymnosperm plants, such as conifers, whose seeds are not contained in an ovary, seem to avoid parthenogenesis totally, while aphids and water fleas alternate between male-female sex and parthenogenesis. The evidence suggests that no major group of organisms (except, possibly, one group of invertebrate rotifers) has evolved and diversified over long periods using parthenogenesis alone.

Two different arguments have been put forward to explain this persistent non-competitiveness of asexual breeding. According to the first, sexual reproduction seems to win out in the long term because the fusion of gametes from different individuals and the consequent genetic shuffling continually throw up sufficient variation within a species to enable it to overcome novel environmental problems by adaptive change.

The various effects of mutations in individual organisms can help to define the advantages of sexual reproduction. A mutation takes place when the make-up of a single gene is changed. Imagine that two beneficial mutations resulting in greater fertility appear in different individuals in the same population of organisms. Each mutation will gradually increase in frequency through time under the influence of natural selection.

In a sexually reproducing population both mutations can be combined in one individual by mating. This combination will happen in an asexual population only when one mutation comes about spontaneously in an individual which already possesses the other. As a result, the combination of beneficial mutations will occur more quickly with sex than without it. In short, groups reproducing sexually can evolve more quickly than those that do not.

The second argument in favour of sex suggests that disadvantageous mutations (rather than lethal ones) may accumulate more easily in asexual than in sexual reproducers. Imagine that only a small proportion of a population lack troublesome mutations such as an inability to digest a particular food. If, in one generation, these individuals failed to reproduce then that proportion would fall to zero. An asexual population may not be able to reverse this decline in overall fitness, but a sexual population could do so through genetic shuffling in a subsequent generation. It is even possible that the proportion of disadvantageous mutations in parthenogenetic populations would progressively increase over many generations.

These two insights – the faster rate of evolution in sexual reproducers and the clogging up of parthenogenetic organisms with deleterious mutations – may help to explain a fact of the living world: if you are a eukaryotic, multicellular organism, sex rules.

For prokaryotes the situation is more complex. True sexual reproduction does not occur in bacteria or blue-green algae. But bacteria, in particular, are now known to use a range of mechanisms for transferring genes from one single-celled organism to another in a kind of pseudo-sex (pp. 188–91).

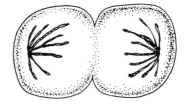

Simple cell division – mitosis – is a form of asexual reproduction employed by single-celled organisms such as protozoa and some algae.

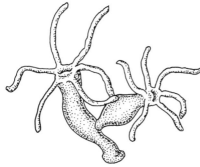

Asexual budding is common in simple animals such as the hydra. Multiple cell divisions build a tiny version of the parent on its side which eventually drops off.

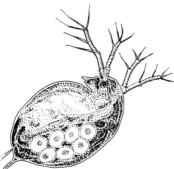

Parthenogenesis, or virgin birth, is used by some crustaceans, such as the water flea *Daphnia*. Non-fertilized eggs can develop into new *Daphnia*.

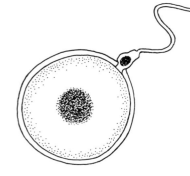

In orthodox sexual reproduction a haploid sperm fuses with a haploid egg to form a diploid fertilized egg. This then develops into a new individual.

Environment as well as genetics influences the sex of offspring. The Nile crocodile, for example, buries her eggs and guards them while they incubate. The temperature during incubation determines whether the eggs hatch into males or females. At 86°F (30°C), all hatch as females; but at 91°F (33°C) all are males. The mother transports her tiny young in her own mouth.

RAW MATERIALS OF EVOLUTION

Mutations have had a bad press. They are often associated with the mutants of horror stories or science fiction, or else with the dangers of atomic radiation. But mutations in the world of living things and on the stage of life's history on Earth are in fact the raw material of evolutionary change.

A mutation is simply an identifiable change in the genetic constitution of an individual, an alteration in the sequence of nucleotide bases on the genome's DNA (pp. 60–61). This is the key to evolution, for mutations represent the intrinsic variation upon which natural selection operates. Some mutations may indeed make the twisted monsters of science fiction. Others, though, became the mechanism by which fish fins turned into amphibian legs, reptilian scales became bird feathers and humans ended up with a larger, more complex brain than their primate ancestors.

Geneticists have found that mutations are triggered in several ways: by radiation, by reactive chemicals known as mutagens or by events related to the copying and construction of DNA.

Mutations are grouped into four

Down feather

Pheasant body feather

Waxwing feather

Barn owl first wing primary

Bird feathers must have evolved from mutated versions of earlier reptilian scales. Subsequent mutations have produced the vast diversity seen in today's bird species.

THE SICKLE CELL GENE

Sickle cell anaemia is caused by an inherited gene that directs the production of human haemoglobin. The mutant gene, known as *HbS*, produces a form of haemoglobin that differs from normal human haemoglobin in just a single amino acid. This minute change, however, produces a cascade of alterations in people with the gene. Some of the red blood cells containing the changed haemoglobin alter their shape and become spiky and irregular – the so-called sickle cells.

These altered cells, with their tendency to clump together, work less efficiently and can cause moderate disease in people with one *HbS* gene and one normal gene or severe illness, and even death, in people with two *HbS* genes.

The *HbS* gene is particularly widespread in Africa, despite its seemingly harmful effects. This apparent paradox is due to the fact that it offers considerable protection against malaria – an often lethal disease also common in Africa. This protection means that there is positive selection for the sickle cell gene in areas where malaria is prevalent (see maps). The advantages of the protection outweigh the disadvantages of the illness.

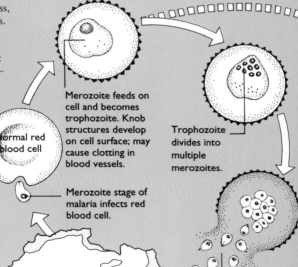

Normal red blood cell

Merozoite feeds on cell and becomes trophozoite. Knob structures develop on cell surface; may cause clotting in blood vessels.

Trophozoite divides into multiple merozoites.

Merozoite stage of malaria infects red blood cell.

Merozoites escape from ruptured red blood cell.

The red blood cells of a carrier of the sickle cell gene are visible in this false-colour scanning electron microscope picture. The elongate spiky cells are sickled and the rounded cell is normal.

Distribution of falciparum malaria in Africa

major types. First is base substitution: one nucleotide base in a DNA sequence is replaced by another. This may have a variety of consequences in a gene that codes for a protein. If the meaning of the gene's triplet code is changed then one amino acid will be replaced by another in the resulting protein. This is what happens when the normal haemoglobin gene mutates into the gene that produces sickle cell anaemia.

Some triplet codes (up to six) may be synonymous in the sense that they all code for the same amino acid. Substitution mutations that lead to synonymous triplets change the DNA but do not alter the protein product, which means they are functionally neutral and play little part in the process of evolution. With a non-synonymous mutation, the effects of a single amino acid change range from lethal to insignificant to beneficial, depending on the importance and role of the amino acid concerned.

The second type of mutation involves a change in a DNA sequence by the insertion or deletion of a nucleotide base. Here the mutation's effects are much more dramatic because such a change induces a "frame shift" in the reading of the triplet code. An insertion or deletion of a base will alter every triplet "downstream" of the change: the triplets will be read starting one base forward or back from the original configuration. This almost always destroys a cell's capacity to make the relevant protein.

The two other mutational types involve alterations to DNA sequences that are longer than one nucleotide base. In the first, a part of the sequence is chopped out and re-inserted in an inverted order. In the second, a whole sequence is duplicated or deleted. The influence that these changes have on an organism's phenotype (pp. 66–67) depends on their length and their position on a chromosome. An inverted section within a functional, protein-coding gene, for example, will almost certainly result in a faulty protein. However, the effects of an inverted or duplicated section in a non-coding sequence between genes will be far less damaging because the functional parts of the genes will remain intact.

Painstaking experiments with such organisms as the bacterium *E. coli* and the fruit fly *Drosophila* have enabled geneticists to calculate how often the mutational types occur. The frequency of base substitution mutation in *E. coli* is, in fact, fantastically low – about once in every billion times a DNA sequence is copied. This remarkable level of copying accuracy results from a series of error-checking steps in the making of each DNA strand. No single step is as accurate as the total error rate suggests, but at each step in the copying process, errors in the previous step are sought and corrected.

This one in a billion alteration, along with the tiny but crucial effects of other mutational types, provides the foundation for evolutionary change.

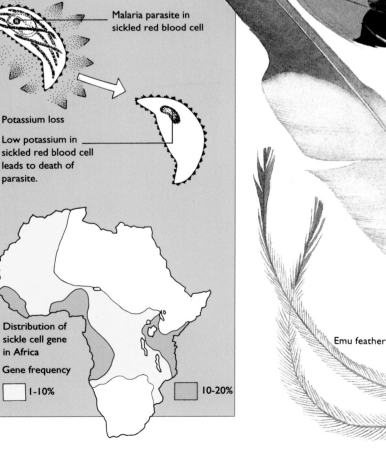

Malaria parasite in sickled red blood cell

Potassium loss

Low potassium in sickled red blood cell leads to death of parasite.

Distribution of sickle cell gene in Africa

Gene frequency

| | 1-10% | | 10-20% |

Motmot tail feather

Buzzard second wing primary

Emu feather

Chicken feather

Penguin feather

NATURAL SELECTION

Evolution's success stories – and its eventual failures – have come about because of natural selection. An individual organism, equipped with new characteristics brought about by mutations in its DNA, needs to be tried and tested. If it proves to be fit enough to survive and reproduce faster than pre-existing types then the new characteristics, and the mutations underpinning them, will spread through populations of the species via the agency of natural selection.

Darwin's ideas for selection came from the artificial process that breeders use to develop a heavier-eared wheat or a leaner pig. Today, geneticists in the laboratory can mimic natural selection. They can experiment, as they have done for decades, with a particular characteristic of a rapidly reproducing organism, such as the number of bristles on a fruit fly.

But such selection operates all the time in the wild, without a geneticist to decide which individual will or will not breed. One species that clearly demonstrates this is Britain's peppered moth *Biston betularia*, which exists in two forms, one speckled, one black. The speckled form was more common until the 19th century, when the black form took over in industrialized areas. More recently, numbers of the speckled form of peppered moth have increased.

When resting on light-coloured lichen on tree trunks, only the speckled form of the moth is camouflaged from attack by insect-eating birds. When soot in the smoke from homes and factories killed the lichens and turned the tree trunks black, it was the black moth's turn to be more efficiently camouflaged.

In this instance, birds are the agents of natural selection. In the 19th century, they ate more speckled moths than black moths – they were easier to spot. The latter bred more, thus altering the genetic structure of the moth's population. But with modern regulations on soot emission and the scaling down of industry the reverse is occurring and the speckled moths are returning.

Sceptics, from Darwin's time until the present, have accepted small-scale, "microevolutionary" changes, such as fruit fly bristle numbers and moth wing

The snowy owl is a flying predator in Arctic regions. Its pure white feathers provide valuable camouflage in the ice and snows of its habitat.

colour. But they baulk at the idea that aggregates of tiny adaptive changes can generate complex organ systems.

The vertebrate eye encapsulates the essence of the dispute. This complex piece of organic engineering blends interacting optical, structural and nervous components. How, the doubters question, could the intricate eye have evolved through mutational variation and natural selection?

A comparison of the components of this eye with the simpler eyes existing in the animal kingdom shows that the organ's complexities could have arisen in developmental stages, none of which need have been impossibly profound.

The eye might have started out as a depression in the skin with light-sensitive receptor cells at the bottom. Flatworms have such simple eyes as these. The depression could have become like a pinhole camera by deepening into a spherical enclosure with a tiny opening to the outside world. *Nautilus*, the primitive cephalopod, has exactly this kind of eye to improve its ability to sense the direction of objects.

A mutation that created a jellylike filling in the sphere could have acted as a crude lens, improving the eye's image-

forming capacity. Some marine annelid worms possess such an eye. Once this stage had been reached, the processes could have been extended to create a denser, more perfectly formed lens; a light-controlling diaphragm, or iris; and, eventually, fine-tuned receptor cells for improved resolving power.

The precise details are not crucial. What is important is that a complex organ could have evolved via myriad intermediary forms. Each was functional, with a small adaptive advantage over the previous engineering design. This is a vital conclusion for it encourages fearless advocacy of natural selection as a mechanism for adaptive change.

In its Arctic home, the polar bear has been subject to the same selection pressures for body colour as the snowy owl. The polar bear's white camouflage gives it a few seconds' advantage when creeping up on prey such as seals.

The flightless cormorant – a large fish-eating seabird that catches its prey under water – is the only member of its family that does not fly. Like some other birds that live on remote oceanic islands, it has lost the power of flight. Since there are no natural predators on the Galápagos and there is a plentiful supply of fish immediately off shore, loss of flight seems to have been a non-harmful trait. It can even be seen as beneficial in that the cormorant has been able to dispense with an extremely energy-expensive activity.

The diagram below shows the direction of the selection pressure on wing length that must have been part of the evolutionary history of the Galápagos cormorant.

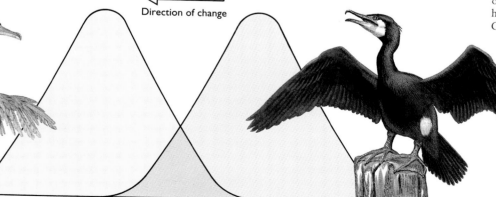

Numbers of individuals

Direction of change

Flightless cormorant

Increasing wing length

Cormorant

GENES AND VARIATIONS

"Like begets like" is the simplest description of reproduction within a species. Yet it is obvious that, apart from identical twins, all members of a species are not alike, not least because they are genetically unique. Gross morphological features of a species, such as the number and form of its limbs, may not vary, but a close look at its individual members will show how they differ from one another.

This inbuilt variation is not always hard to see; we make use of it each time we recognize a person we know. Equally, creatures such as the land snails of the genus *Cepaea* show startling variations of shell colour and banding pattern within a single species. The species is said to display polymorphism: its members exist in a great variety of forms.

A true measure of the polymorphism in a species will, ultimately, only be possible by reading the DNA sequences of its genome. For the present, a good idea of underlying genetic variation can be gained by assessing the protein products of a species' genes.

A quick and simple laboratory technique which differentiates most proteins is gel electrophoresis. It measures the net positive or negative charge on a protein molecule. A protein is made up of a specific sequence of amino acids, each with its own charge. Since different proteins have different mixes of the 20 possible amino acids, the total net charge on each one will usually be different. Even a single amino acid change between one protein and another may produce an alteration in overall charge.

Slight alterations in the amino acid sequence of a protein do not impair its general function. Instead, they give it variety. Electrophoresis is used to assess the number of forms that a protein can assume. It is less accurate than charting the amino acid sequences of every form of the protein but it is much faster.

When electrophoresis was first applied to proteins from humans and fruit flies in the 1960s, the results were astounding. Between a quarter and a half of the protein types examined were polymorphic – they existed in more than one charged form and therefore must be represented by a different gene version.

For example, human haemoglobin has more than a hundred variants.

It would be reasonable to think that one of the protein variants of each gene would be of a "higher fitness" than the others and therefore, by natural selection, become more and more common to the exclusion of all others. But this monomorphism does not always occur

and the mechanisms which maintain the diversity are the subject of controversy among evolutionists.

One mechanism which may conserve a suboptimal gene is balanced polymorphism. A gene that produces a suboptimal protein may persist in a population because, in tandem with another gene, it provides a clear fitness advantage. For

Individual land snails of the species *Cepaea nemoralis* display incredible variation in their colour and banding patterns. But despite differences in these genetically determined characteristics, the snails are all still members of the same species. Some variations may give a population of snails a selective advantage in their particular habitat. For example, pale shells may help snails on light-coloured soils stay camouflaged from snail-eating birds.

Humans all belong to one species but within that species there is great variation. Since we are spread over the globe in many climatic zones, a range of characteristics such as different skin colours and hair textures and colours have evolved which confer advantages in certain conditions.

example, the human sickle cell haemoglobin gene persists in Africa despite its harmful side effects because, in individuals who possess one normal and one mutant gene, it protects against malaria.

A second reason relates to the changing selective pressures that exist from area to area and from time to time. This may mean that no specific adaptive choice between locally advantageous variants ever occurs in a population.

The Japanese geneticist Kimura has proposed a third possibility. His "neutral theory" suggests that, in terms of selective advantage, protein forms are neutral with respect to one another. They do not make the individuals carrying them any more or less successful than those

carrying other variants. This may be why so many forms of a protein exist simultaneously and transiently.

Whatever maintains the large degrees of polymorphism found in a species, the variation itself is more than the raw material of adaptive change. It is the first step on a hierarchy of difference which leads to the formation of a new type of organism – a new species.

But what is a species? In everyday language, the word describes a sort or type of living creature. For biologists, and more particularly evolutionary scientists, drawing the line between one species and another poses enormous questions. What defines the boundary of a species in the first place and how is a new species formed?

Organisms that are clearly distinguishable from one another, such as an orangutan and a chimpanzee, are easily labelled as different species. It is as though they each belong to a group which has obvious boundaries. Recognition of these boundaries appears to be widespread. The Kalem tribe in New Guinea, for instance, recognize and have special names for 174 types of forest-living vertebrates – reptiles, mammals and birds. All except four of these names correspond to the scientifically defined species recognized by biologists.

The species is, arguably, the only absolutely definable and therefore "natural" grouping in the hierarchy of the classification system (see box). This is because the organisms constituting a species define the group they belong to by their own observable behaviour.

"A species", to quote the evolutionary zoologist Ernst Mayr, "is a group of actually or potentially interbreeding populations that are reproductively isolated from other such groups." Put simply, this means that the members of one species can breed with each other (and produce viable offspring) but they cannot breed with members of other species. Thus the breeding behaviour of organisms explicitly describes the boundaries of a species. Obviously, this definition only applies to creatures that breed sexually. Asexual reproducers may have identifiable types called species but these cannot be properly defined by Mayr's generalization.

In practice, breeding behaviour is the primary characteristic that determines where a species boundary lies. Secondary to this, scientists further define a species using a range of physical characteristics, such as numbers and types of animal teeth or the numbers and shapes of flower petals, which can all be used as a checklist for identification purposes.

Deciding when and how a new species has evolved presents greater problems than defining an existing species. The formation of a new species, or speciation, may involve the complete transformation of one species into another (anagenesis) or the splitting up of one species into a number of others (cladogenesis). Whichever form it takes, speciation is the direct result of changes in the gene pool, and therefore the genetic polymorphism (pp. 78–79), of a species.

When all the populations of a species live in the same geographical area they are said to be sympatric – that is, in continuous contact with one another. Consequently, genes can flow with relative ease throughout the total gene pool of the species. A beneficial mutation in a gene in one population can lead to an adaptation that spreads throughout the pool by natural selection over many generations of breeding.

With genetic continuity of this kind, a species is unlikely to split into a number of others. Genetic differences between individuals in one part of a species' range and another could be diluted by interbreeding and gene flow, so reducing the chances of speciation.

What is possible, though, is the transformation of one species into another. Adaptations within a species over a considerable time can result in such dramatic changes to physical characteristics that it is no longer the species it once was and therefore has to be given a new species name. This kind of speciation incurs no disruption to genetic continuity and

SPECIES SPLITTING

Occasionally, one species does split into two without geographical separation as has apparently happened with these two North American lacewing species. Only a few genes distinguish the two, but these tiny genetic variations can have far-reaching effects.

The widespread lacewing species, *Chrysopa carnea*, lives in deciduous woodland and is pale green in summer and brown in winter. The new species, *Chrysopa downesi*, evidently derived from *C. carnea*, lives among evergreen conifers and is dark green all year round. The crucial factor is that the two breed at different times of the year (see diagram) and so are never in breeding condition at the same time. Gene flow between the two is therefore impossible.

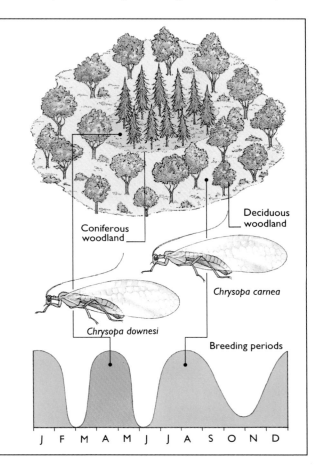

Coniferous woodland

Deciduous woodland

Chrysopa carnea

Chrysopa downesi

Breeding periods

J F M A M J J A S O N D

A species is a group of individual organisms that can breed with one another. Each has a two-part (binomial) name in Latin which can be used universally for that organism. First is the generic name – that of the genus to which that organism belongs. Second is the specific name – the particular and sometimes descriptive name identifying that one species.

Species are grouped into genera, genera into families, families into orders, orders into classes, classes into phyla and phyla into kingdoms in a hierarchy of ever more inclusive levels.

The levels of classification of one species, the lynx, *Felis lynx*, are shown here.
Genus *Felis*: a group of closely related cats, distinguished by behaviour, habitat and coloration.
Family Felidae: meat-eating cats with five digits on front paws, four on hind paws.
Order Carnivora: mostly flesh-eating animals, including cats, dogs and bears.
Class Mammalia: warm-blooded animals which suckle their young on milk produced by mammary glands.
Phylum Chordata: animals with a notochord (forerunner of the backbone) above which is a hollow nerve cord.
Kingdom Animalia: all animal forms.

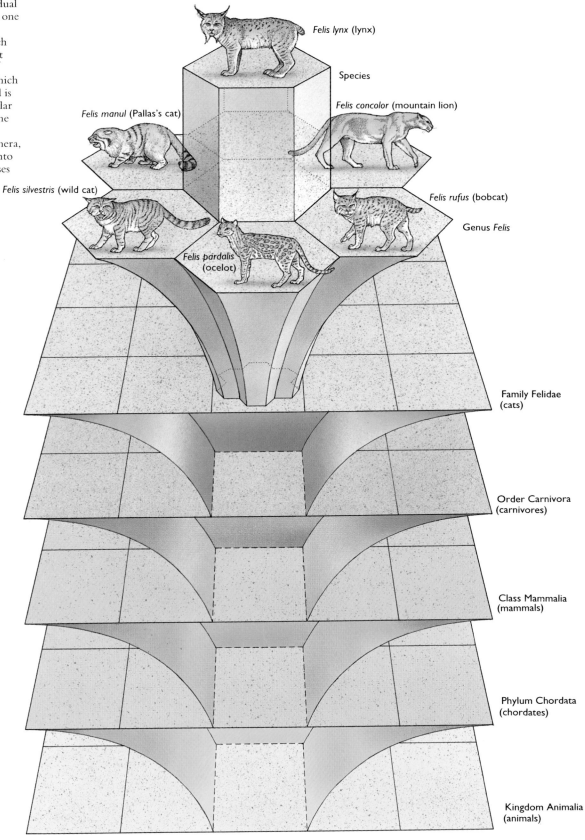

Felis lynx (lynx)

Species

Felis concolor (mountain lion)

Felis manul (Pallas's cat)

Felis silvestris (wild cat)

Felis rufus (bobcat)

Genus *Felis*

Felis pardalis (ocelot)

Family Felidae (cats)

Order Carnivora (carnivores)

Class Mammalia (mammals)

Phylum Chordata (chordates)

Kingdom Animalia (animals)

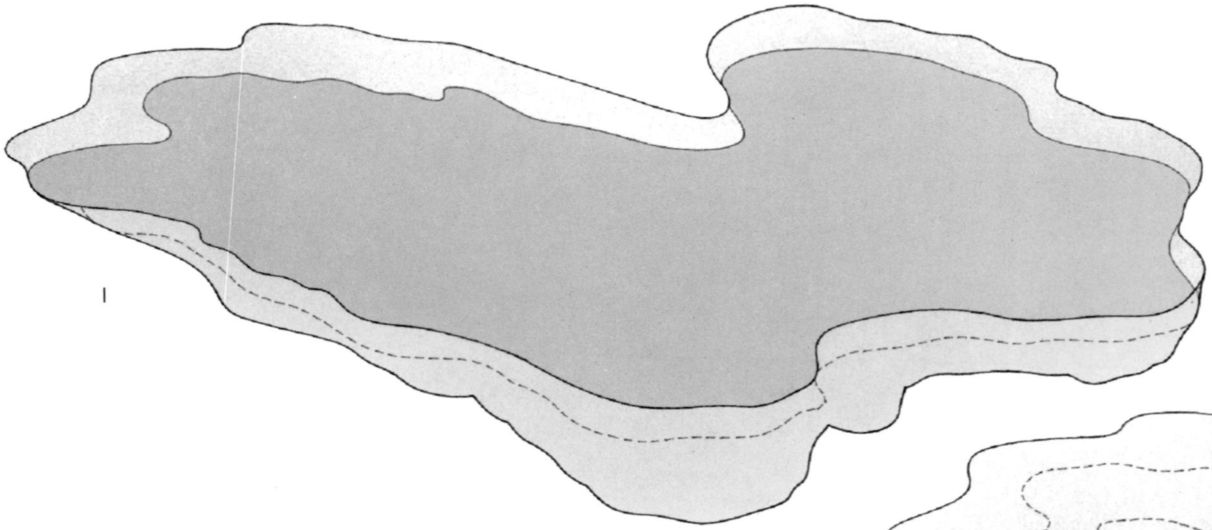

1 **The diversity of cichlid species** in the African lakes could have been caused by changes in water level over thousands of years. The first stage in this hypothetical sequence shows the lake at its fullest, with breeding possible between all populations of a species.

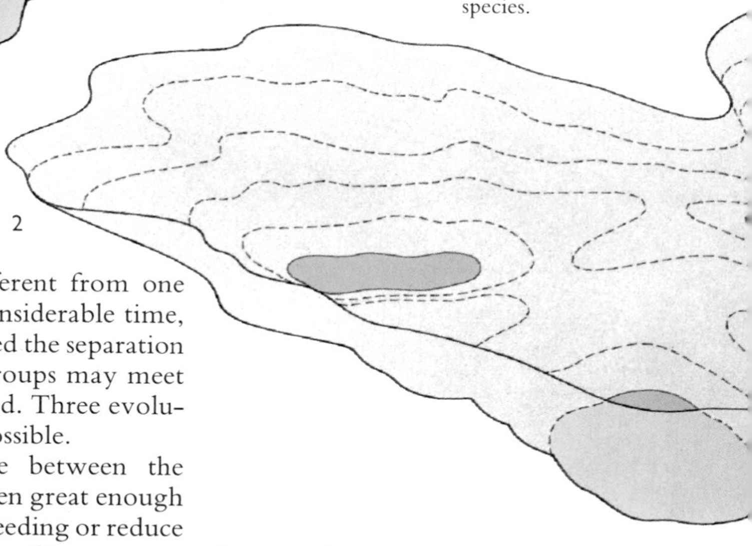

the members of the new species have an unbroken lineage from those of the species that preceded it.

Species splitting, a central part of the dynamics of evolution, is probably the most important way new species are formed. Such speciation usually occurs when populations of a species come to live in different geographical areas. The process, known as allopatric speciation, occurs when persistent genetic differences between the populations result in geographical subspecies which eventually become new species.

The basic idea is relatively simple. In any widespread species, local genetic variation produces identifiable, different-looking subspecies, such as plants with varying petal markings on their flowers or birds with different plumage patterns. But the species as a whole is kept intact because of gene flow between the subspecies. If, however, a climatic, geographical or geological alteration produces an insuperable barrier to breeding between such groups, genetic continuity between one population and the rest will be prevented. Advancing ice sheets, changing river courses, falling lake or sea levels or expanding deserts can all produce physical barriers that split populations. The sequence illustrated here shows the possible consequences of a fall in the water level of a lake.

Once divided, the two populations will develop genetically under the influence of independent selective pressure brought on by differing environ-

ments and become different from one another. But if, after considerable time, the changes that produced the separation are reversed, the two groups may meet again and may interbreed. Three evolutionary outcomes are possible.

First, the divergence between the groups may not have been great enough either to prevent interbreeding or reduce significantly the viability of inter-group hybrids. As a result, the groups will merge back into one species and gene flow will resume.

Second, the divergence may be great enough to prevent viable interbreeding. The two groups may, perhaps, be ecologically isolated. If plants, they may flower at different seasons or be pollinated by different insects. If animals, they could be faced with differences in courtship behaviour or in mate recognition. The lack of gene flow between them means that they must be called two different species. They might be able to coexist or one might be so much more successful that the other becomes extinct.

Third, an intermediate situation may result from the resumption of contact. Thought by many evolutionists to be a common circumstance, this carries with it the seeds of further consolidation of incipient new species groups.

The implication of this scenario is that genetic differentiation in isolation goes

far enough to produce some changes but not to prevent all interbreeding. If the offspring of within-group matings are more successful in survival and fertility terms than those of between-group matings, strong selection pressures will act against the latter.

Closely related species, especially those that probably arose from a split in the recent past (so-called sibling species), may look the same yet possess some key features that preclude hybrid matings. In Britain, for instance, the three common leaf warblers, wood warbler, chiffchaff and willow warbler, are remarkably similar in both size and plumage.

These sibling species, however, have profoundly different mating season songs. Since song recognition is a key component of pair-bonding between males and females, different songs will make cross-group mating attempts unlikely. By such isolating mechanisms, incipient species differences are made permanent by the closing down of the gene flow between populations.

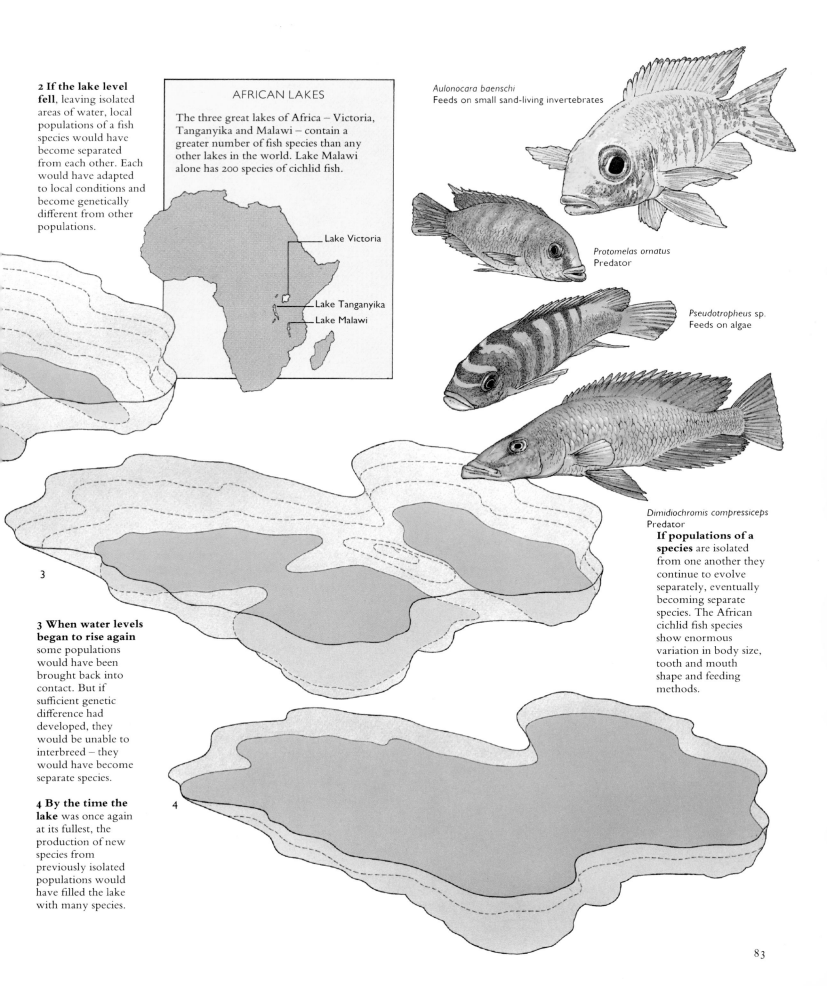

2 If the lake level fell, leaving isolated areas of water, local populations of a fish species would have become separated from each other. Each would have adapted to local conditions and become genetically different from other populations.

AFRICAN LAKES

The three great lakes of Africa – Victoria, Tanganyika and Malawi – contain a greater number of fish species than any other lakes in the world. Lake Malawi alone has 200 species of cichlid fish.

Lake Victoria

Lake Tanganyika

Lake Malawi

Aulonocara baenschi
Feeds on small sand-living invertebrates

Protomelas ornatus
Predator

Pseudotropheus sp.
Feeds on algae

Dimidiochromis compressiceps
Predator

If populations of a species are isolated from one another they continue to evolve separately, eventually becoming separate species. The African cichlid fish species show enormous variation in body size, tooth and mouth shape and feeding methods.

3

3 When water levels began to rise again some populations would have been brought back into contact. But if sufficient genetic difference had developed, they would be unable to interbreed – they would have become separate species.

4 By the time the lake was once again at its fullest, the production of new species from previously isolated populations would have filled the lake with many species.

4

CHANGE AT A HIGH LEVEL

Evolution resembles a great epic with a cast of millions. Organisms past and present play their part in a lengthy saga which charts the creation of new species. The epic, though, is a play without a script – evolution has been 4 billion years of improvisation.

Organisms are classified into species, genera, families, orders, classes and phyla according to shared characteristics. Each of these groups, or taxa, contains more types than the one preceding it and is defined by characteristics that are more and more general. Each level of this classification, this hierarchy of difference between organisms, pinpoints for any particular group a phase in the continuous creation of new species. In other words, it represents a moment in evolutionary time when a significant set of changes was established.

The major differences in body plan which distinguish one phylum from another – for example, arthropods from chordates – are statements of the changes that became established in the distant past. The fossil record makes it clear that most of the invertebrate animal phyla that are alive today were already in existence by the end of the Cambrian 500 million years ago, if not before.

The Cambrian seas spawned a wealth of eukaryotic animals with different body plans which can be grouped into phyla: coelenterates, annelids, molluscs, echinoderms, priapulids, lampshells, arthropods, and probably many others, including invertebrate chordates.

Because this differentiation occurred so long ago, the true lineages of the phyla cannot be tracked easily or directly. DNA evidence from living representatives of these groups provides us with some clues, but much of our description of this crucial phase of evolutionary history is based on informed speculation rather than hard fact.

The first key question to be answered is how did single eukaryotic cells evolve into more complex, multicellular invertebrates? Such a change could have come about by a sequence of small developments which could have produced significant alterations in body plans.

Suppose such a single cell lived like a modern-day protozoan. The first development would have had to involve mutations which prevented the cells from moving apart after they had made copies of themselves by the process of mitosis.

These mutations presumably activated one or two key proteins in the cell membrane that enabled the surfaces of daughter cells to stick together. Multiple divisions would generate genetically identical cells as either a solid mass or two-dimensional sheets.

The inner and outer regions of the cell mass would have experienced slightly different environmental conditions – for instance, oxygen would be more available on the surface than internally. Such differential stimuli could have led to the activation or suppression of particular genes among the genetically identical cells. If this meant that surface cells had externally directed flagella or cilia but internal ones did not, a recognizable change of body plan would have resulted. This is the so-called planula body architecture, named after the larval stages of coelenterates of this type.

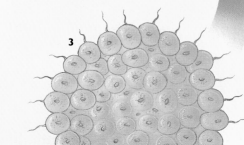

Simple gut

Flatwormlike organism

4

Simple many-celled organism

3

Many-celled animals are believed to have developed from single-celled ancestors such as amoeba-like protozoans (1) or forms bearing tail-like flagella (2). In the late Precambrian several lines of multicellular development may have arisen when single cells divided but did not separate, thus producing cell clusters.

The simplest clusters were solid, probably with flagellated cells outside and non-flagellated cells inside (3). Similar animal forms still exist today, such as placozoans and mesozoans (*above left*) and the planula larvae of coelenterates.

The crucial next stage of development was the intucking of the outer surface to form a simple gut without an anus (4). Such a simple body plan, similar to that of today's flatworms, probably gave rise to most other types of invertebrate groups.

Single-celled protozoans

1

Flagellum

2

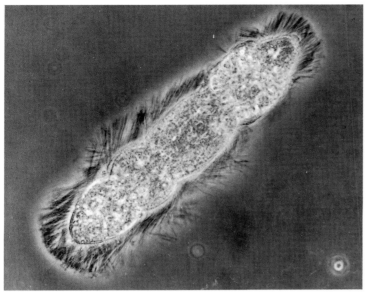

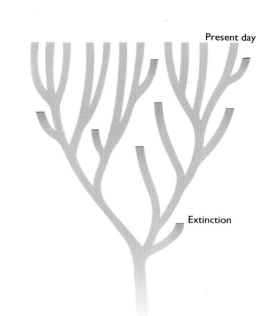

These two male mesozoans are simple animals, with no gut, blood system, muscles or nervous system, which live as parasites inside a brittle star. Their bodies are solid cell masses with the outer layer of cells bearing flagella.

Some present-day phyla of minute invertebrates have planula body plans — no nervous system, no blood system, no gut, simply a flagellated or ciliated outer rind of cells and non-flagellated cells internally. The free-living placozoans are built this way, as are the parasitic mesozoans (see above).

Only one major mutation would then have been needed to transform the solid, regionally differentiated mass into an organism with a flatworm body plan. In its simplest form, this plan involves the outer surface cells tucking into the centre of the mass to produce a basic sac-shaped gut. This ancestral flatworm is significant because, according to many evolutionists, it could have been the stem group from which many other, more complex, invertebrate phyla arose.

Each mutation in the sequence that turned the single cell into a simple flatworm must have had selective advantages in the competitive Precambrian environments. No one knows what those advantages were. Perhaps cell groups could engulf larger food items; locomotory cells on the surface of a cell mass may have brought more widespread food sources within reach; and the pocket-shaped intucking of the flatworm must have made food trapping and digestion more efficient.

The "shape" of the evolutionary tree of life is difficult to construct because of the incomplete nature of the fossil record. From the material available, however, one view suggests that there is a "cone of diversity" with an ever-increasing range of body plans arising from restricted beginnings (*above*). This widening range has continued through time, despite random extinctions. A more complex theory suggests that for a considerable period a high diversity of organisms has arisen from a range of origins (*below*). In this "pruned bush" scheme, mass extinctions periodically remove a large proportion of organisms, leaving the remainder to diversify once again.

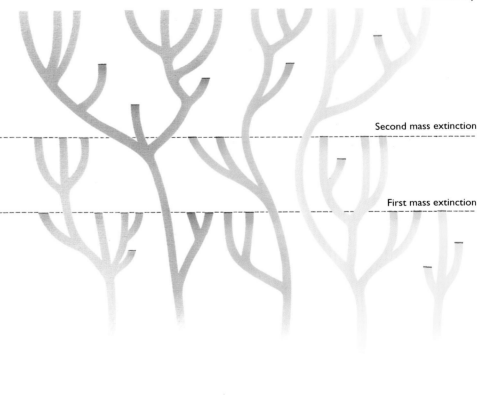

85

VARIATIONS ON BODY PLANS

A fundamental tenet of evolution rests upon life's ability to generate new designs from pre-existing components. The design features that typify a phylum often result in remarkably diverse organisms which not only look very different but also live very different lifestyles. In extreme instances, selection may force some of these features to be abandoned. No typifying feature, therefore, is sacrosanct in the face of evolutionary change – everything is modifiable. Every mollusc species without a shell, every flightless bird is a testament to this fact.

The soft-bodied molluscs are an ancient multicellular invertebrate phylum. The earliest rock strata of the Cambrian period harbour fossils of shells that clearly belong to this group. Thus, their real beginnings must be sought in the late Precambrian.

The members of the mollusc phylum illustrate an extraordinary diversity of forms which, to the uninitiated, seem very unalike. This single phylum contains bivalves (shellfish, such as clams, oysters and mussels), gastropods (snails and their allies, such as limpets and slugs) and cephalopods (octopuses, squids and the now extinct ammonites and nautiloids.

All these groups are very different from one another. It is hard to imagine that a shelled, static mussel tied to a rock by byssus threads belongs to the same phylum as a streamlined, fast-swimming, intelligent squid. And the total diversity is all the more remarkable when the more obscure classes of mollusc – the long, conical tusk shells (scaphopods), the eight-shelled chitons (polyplacophorans), the limpet-like monoplacophorans and the wormlike aplacophorans – are added to the list.

Despite this evident variation, close inspection reveals that the major patterns of evolutionary change involve alterations of a single pre-existing basic body plan. This plan, believed now to be one derived early on from ancestors resembling flatworms (pp. 84–85), has a through-gut, with mouth and anus, no evidence of real segmentation, a calcareous shell, a rasping, filelike feeding organ called a radula and a pocket-shaped space (the mantle cavity) which

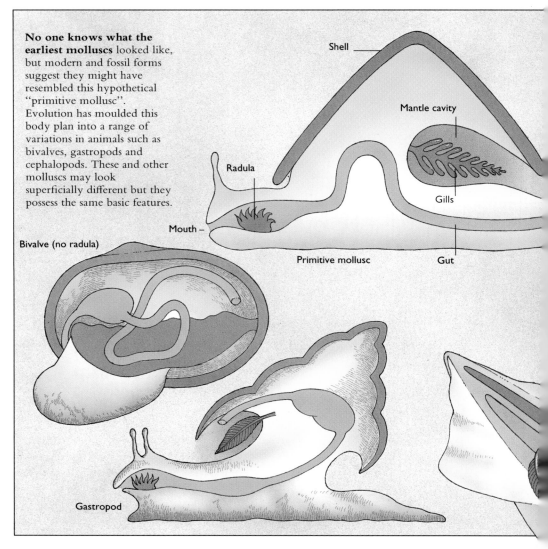

No one knows what the earliest molluscs looked like, but modern and fossil forms suggest they might have resembled this hypothetical "primitive mollusc". Evolution has moulded this body plan into a range of variations in animals such as bivalves, gastropods and cephalopods. These and other molluscs may look superficially different but they possess the same basic features.

Shell

Mantle cavity

Radula

Gills

Mouth

Primitive mollusc

Gut

Bivalve (no radula)

Gastropod

contains gills.

During evolution, the importance of individual key structural elements has been modified, upgraded or demoted in the different molluscan groups. Only in rare examples have they lost these elements or have their functional equivalents been lost completely.

In almost all molluscs gills are used entirely for acquiring dissolved oxygen from water. The bivalves, however, also use them as an extensive, folding, filter-feeding net. This single change enables the bivalve shellfish to benefit from a new set of niches and food sources inaccessible to other molluscs.

The chordate phylum, to which humans belong, is arguably more impressive in its adaptive variation than

the molluscs. The basic chordate body plan includes a segmented body; a hollow main nerve cord along the back; holes or slits in the side of the foregut (the pharyngeal slits) which lead to the outside world; and a strengthening rod, or notochord, which stretches most of the length of the animal.

The notochord must have been the main skeletal support in the earliest swimming vertebrates and has retained this function in sea squirts and in the lancelets (pp. 88–89) In jawless fish such as lampreys, however, the skeletal function has been partially taken over by cartilaginous thickenings around the notochord and nerve chord – these are the beginnings of vertebrae. In higher vertebrates, fish, amphibians, reptiles,

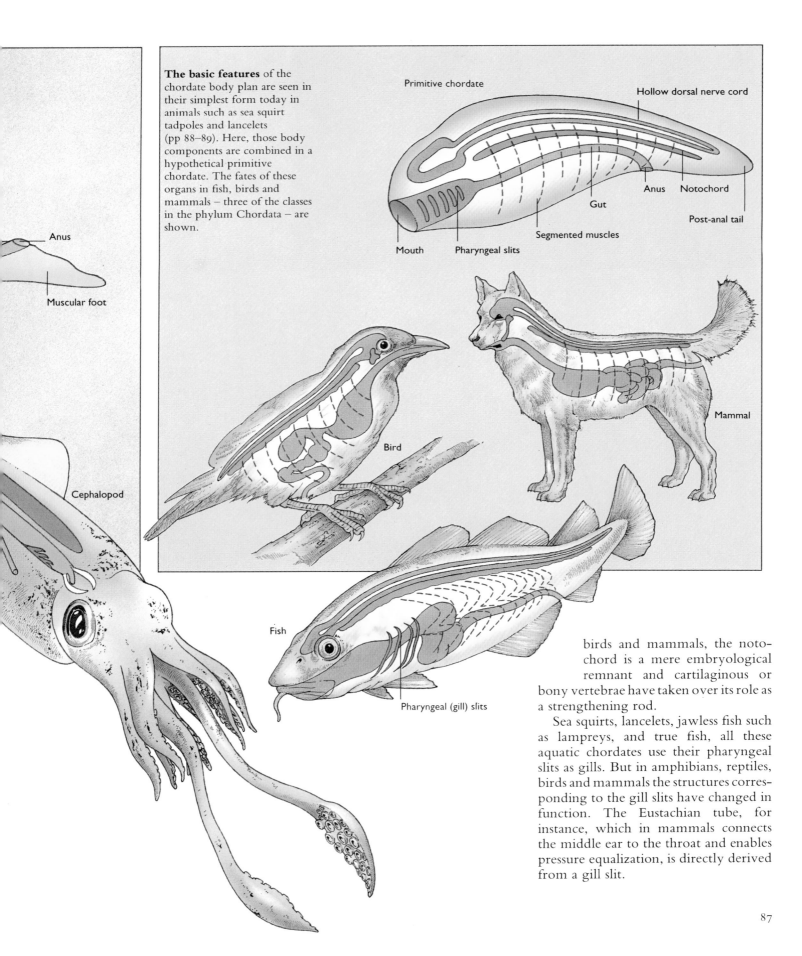

Anus

Muscular foot

Cephalopod

The basic features of the chordate body plan are seen in their simplest form today in animals such as sea squirt tadpoles and lancelets (pp 88–89). Here, those body components are combined in a hypothetical primitive chordate. The fates of these organs in fish, birds and mammals – three of the classes in the phylum Chordata – are shown.

Primitive chordate

Hollow dorsal nerve cord

Anus   Notochord

Gut

Post-anal tail

Segmented muscles

Mouth   Pharyngeal slits

Bird

Mammal

Fish

Pharyngeal (gill) slits

birds and mammals, the notochord is a mere embryological remnant and cartilaginous or bony vertebrae have taken over its role as a strengthening rod.

Sea squirts, lancelets, jawless fish such as lampreys, and true fish, all these aquatic chordates use their pharyngeal slits as gills. But in amphibians, reptiles, birds and mammals the structures corresponding to the gill slits have changed in function. The Eustachian tube, for instance, which in mammals connects the middle ear to the throat and enables pressure equalization, is directly derived from a gill slit.

GREAT TRANSFORMATIONS

One of evolution's most spectacular triumphs was the transformation of terrestrial reptiles into flighted birds (pp. 24–25). Two other evolutionary "gear changes" that opened up new, expansive ways of life were the transition of fish into amphibians and the possible derivation of swimming chordates from ancestors like sea squirts.

When amphibians colonized the land in the Devonian about 400 million years ago they were the first vertebrates to do so. The origins of the earliest forms, such as *Ichthyostega*, have long puzzled paleontologists, but they do all agree that the early amphibians must have evolved from one of the three groups of lobe-finned (sarcopterygian) fish that existed at the time. These groups are the lungfish and the coelacanths, both of which have living representatives, and the rhipidistians, which are all extinct.

The central axis of many bones which supported the fleshy paired fins of these fish seems an obvious candidate for the progenitor of the limb bones of amphibians. The muscle blocks that moved the fins could easily have been modified through evolution to power walking movements on land. Each of the three fish groups has had its advocates as the ancestral source of land vertebrates – the rhipidistian fin bones are remarkably similar to those of early amphibians, while the development of the lungs and limbs of modern lungfish shows con-

Sea squirts or tunicates (*above and below*) are filter-feeding marine animals that, as adults, live fixed to rocks. They are, in fact, simple chordates – members of the same phylum as all vertebrate animals – but their chordate features are only fully evident in the mobile larval tadpoles. When the larvae settle and become adults, features such as the notochord are lost and the nerve cord is reduced to a small solid ganglion. The only clues to their chordate heritage are the slits in the pharynx wall used for filter feeding. The earliest chordates may have resembled these larval sea squirts.

Nerve ganglion

Perforated pharynx

Adult tunicate

Hollow dorsal nerve cord

Perforated pharynx

Notochord

Larval tunicate

Hollow dorsal nerve cord

Segmented muscles

Perforated pharynx

Tail fin

Lancelet

Anus   Notochord

Mouth

The lancelet is a primitive chordate with a body structure simpler than that of a jawless fish. Like the sea squirt, it lives on the seabed and filter feeds through gill slits in the pharynx. But the lancelet retains its swimming features as an adult and can move around with the aid of its muscular tail and fins.

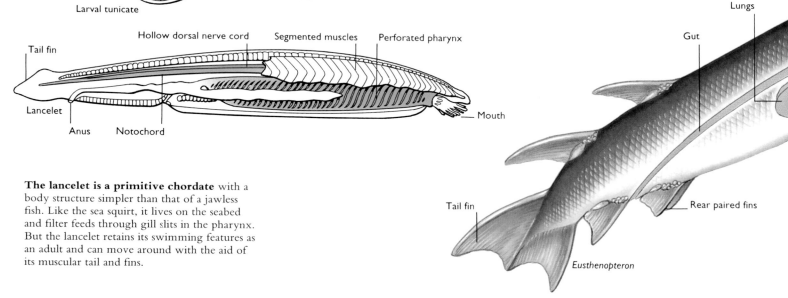

Lungs

Gut

Tail fin

Rear paired fins

Eusthenopteron

siderable similarities to that of living amphibians. All three fish groups are thought to have had lungs in the Devonian.

The fish-to-amphibian transition is illuminated by specific fossil features, whether the layout of bony struts in a fin or the precise positioning of a nostril hole in a palate bone, even though the bones concerned are 400 million years old.

There is no such clear fossil evidence to help pinpoint the origins of the early invertebrate chordates that gave rise to the vertebrates, but living species of simple chordates offer indirect clues. Lancelets, such as *Branchiostoma* (formerly called *Amphioxus*), are minute bottom-dwelling sea creatures that have almost every vertebrate characteristic – except vertebrae. A lancelet swims like a fish with segmented muscles; its body is

stiffened with a notochord; it has gill slits in its pharynx and its nerve cord is dorsal and hollow. Few mutational changes would be needed to convert an ancestral lancelet-like creature into an early jawless fish or proto-vertebrate.

But what are the origins of the lancelet grade of organization? One fascinating theory turns on an unlikely group of chordates, the sea squirts, or tunicates. Adult sea squirts are relatively common on the seashore and in offshore waters below the low-tide mark where they are important members of the community of marine, filter-feeding invertebrates. Sea squirts are tubular animals in which the front or pharyngeal part of the gut has been transformed by many fine holes and slits into a wonderful filter for straining sea water.

Adult sea squirts stay in one place, but

their larvae, known as tadpoles, are adapted for dispersal. Their body form, so different from that of their parents, has prompted taxonomists to place the group within the chordate phylum which contains all vertebrate animals.

Within its elongate body the tadpole develops a flattened tail region which is stiffened by a notochord and powered by segmented muscles. This structure enables the tadpole to swim through water and then settle on the sea bottom. Here it loses its swimming gear and transforms into a filter-feeder that does not move. One evolutionary hypothesis is that swimming chordates, such as the lancelets, arose from those tadpoles that extended their larval life and became sexually mature – a change known as neoteny – before they lost the ability to move through water.

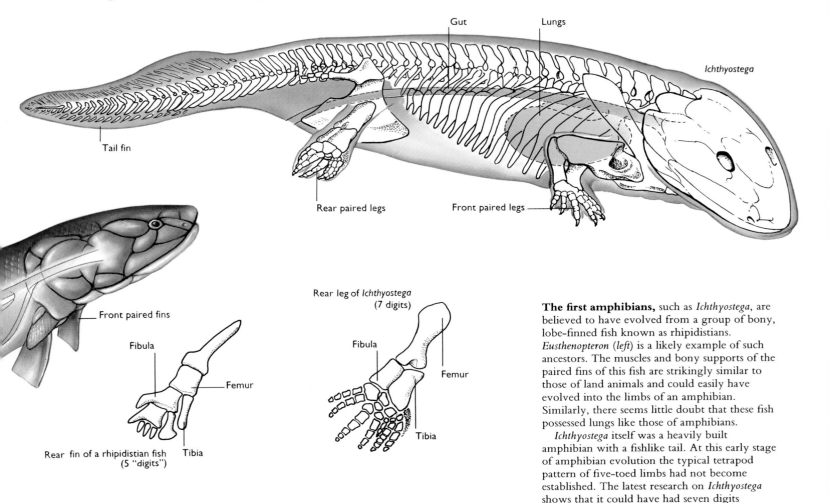

The first amphibians, such as *Ichthyostega*, are believed to have evolved from a group of bony, lobe-finned fish known as rhipidistians. *Eusthenopteron* (*left*) is a likely example of such ancestors. The muscles and bony supports of the paired fins of this fish are strikingly similar to those of land animals and could easily have evolved into the limbs of an amphibian. Similarly, there seems little doubt that these fish possessed lungs like those of amphibians.

Ichthyostega itself was a heavily built amphibian with a fishlike tail. At this early stage of amphibian evolution the typical tetrapod pattern of five-toed limbs had not become established. The latest research on *Ichthyostega* shows that it could have had seven digits supporting each paddle-like hind foot.

LIFE FITS THE WORLD

Genes are the ultimate units of natural selection. From the bewildering variety of altered genes produced by mutations, natural selection sorts out those which contribute to or improve overall reproductive success in particular circumstances. If a slight variation in a gene leads to an adaptation that promotes a greater likelihood of producing offspring then it will be selected in preference to the gene from which it was derived.

An adaptation is a change which enables an organism to operate better and more effectively than a similar organism that lacks the change. A species living in a particular environment, at a particular time and in a community of other species that share an ecosystem, will benefit from an adaptive change. Such changes may take place at any level, from the molecular to social organization, from the sensory to the partnerships shared in coevolutionary contexts.

The machinery of natural selection drives the process of adaptation. Because the ways in which reproductive success can be augmented are almost limitless, adaptive changes occur at every level in the seamless hierarchy of the components and processes of life. However, the molecular prime movers of these changes are effectively all the same. Gene mutations are changes in the DNA nucleotide sequences of an organism's genetic code. Those changed sequences specify slightly altered protein molecules – and everything else in the world of adaptational change stems from these subtly transmogrified protein molecules.

This generalization of the nature of change holds true whether the adaptation is a greater oxygen-carrying capacity in the blood of a goat living at high altitudes, brighter colours in the petals of a flower that must attract insects for pollination, or a distortion in

the shape of an insect's body so that it resembles a twig and is camouflaged. The range of possible changes is so enormous because altered gene-coded proteins can affect almost every aspect of a living creature.

Structure, physiology, biochemistry and development are essentially the province of proteins. These versatile substances are the building blocks of cell components – living structures are literally fashioned from proteins. Proteins are also the enzymes that catalyse all of the thousands of chemical reactions in a cell's metabolism. During cell differentiation and the development of an individual from a fertilized egg, proteins act as feedback controls on genes. Thus, many proteins specifically influence the activity of genes, enabling them to "talk" to one another.

The favouring of changes that are helpful in life's actual context seems to suggest that some guiding force is pushing evolution in a preordained direction. This is a recurring and seductive misconception. There is no external intelligence in life that guides evolutionary change. Evolution is not moving toward a predetermined goal nor is natural selection equipped with a compass. Since evolutionary change is born of random variation and local contingency, the only direction it can take is decided by those randomly generated changes which happen to be successful in specific circumstances.

Some find this basic undirectedness of evolution threatening. They feel the need for an ordering beyond that which natural selection engenders. In fact, though, the extraordinary range, delicacy and intricacy of the adaptations produced, as far as we can judge, by so-called blind evolution is in no way threatening. Viewed in its totality, the range of adaptations of organisms and the web of evolutionary lineages stretching back in time more than 3 billion years are the wonder of the universe. We know of nothing else that is so complex, so interconnected. And it was all made by the process of natural selection.

1

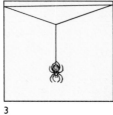

2

3

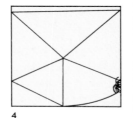

4

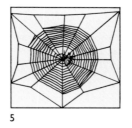

5

A spider builds its web in a sequence of predetermined stages. First it makes a framework of non-sticky threads, firmly attached to the surrounding vegetation (1–4). It then begins to add the radiating spokes of the web (4). When these are all in place the spider adds the central spiral of sticky silk (5) on which flying insects can be trapped.

ADAPTATIONS OF STRUCTURE

A Boeing 747 airliner, a 30-storey office block and a fruit fly are all complex structures pieced together from simpler components according to pre-existing plans. With the blueprint of a 747 in front of them, aviation engineers can fashion an exact copy of the airplane. Given the detailed plans of an office block, an architect can erect the same block elsewhere. Similarly, under the control of the genes on its eight chromosomes, a fertilized fruit fly egg cell grows into a perfect new fly. Of these three examples the tiny fruit fly is by far the most complex.

As in all organisms, the fruit fly exhibits different levels of structure. Individual cells in the fly are specialized to build particular proteins, both for the living parts of the body and, with additional macromolecules synthesized by the cells, for non-living features such as the cuticular scaffolding of the wings. Groups of similar cells are packaged together to make the tissues and organs of the whole organism. This detailed organization is guided by the genes which control all of the fly's development.

The same sort of rules govern the construction of all living things, no matter what their size or shape. Genes direct the building of structures such as a blue whale's immensely long gut, a tiny fruit fly's wing and microscopic cells. Because all the various types of living structures are organized by genetic activity, they are prone to evolutionary selection pressures and can be shaped for survival by the processes of adaptation.

The structure of any organism, whether it be animal, plant or microbe, is the culmination of its evolutionary history, a history of speciation and adaptation. This all-inclusive history takes into account everything directly or indirectly determined by inheritable genes. The full evolutionary tale of a spider, for example, would include the structural adaptations that resulted not only in its ability to spin a web – made of protein silk coded for on spider genes and built according to inherited behavioural rules – but also in its eight hydraulically operated legs and its poison-delivering fangs.

Evolutionary adaptations bring about changes to the form and lifestyle of a species over a long period of time. But there are no general rules that decide which direction evolutionary adaptation takes. The nature of the changes, however, is

The structures an animal makes can evolve
as well as the creature itself. This orb web is made of silk proteins coded for on spider genes. Even the web's pattern is determined by specific web-building behaviour, also genetically determined. Both the composition of the silk and the web design are inheritable. Thus they can – and do – undergo evolutionary change.

totally dependent on two starting points. First, the genotype upon which selection must operate and, second, the specific environmental and competitive context in which the species lives.

A mammal cannot, for example, improve its nutrition by photosynthesizing like plants, although this would be a superb adaptation in sunny climates. It is impossible because mammals do not start with a genotype containing any genes that could conceivably generate photosynthetic molecular machinery.

Some recurring competitive conditions do seem to generate particular trends of adaptational change, such as that for large size in either plants or animals. The evolutionary advantage conferred by large bulk differs according to the species and its ecological position. For carnivorous animals large size can place a species, such as the great white shark, at the pinnacle of a food chain.

For a forest tree, height, size and elevation provide the selective advantage of spreading the light-catching leaves above those of neighbouring species. Such trees thus succeed in the competition for available sunlight.

VASCULAR BUNDLES

The size of such plants, however, poses bioengineering problems for which structural adaptations have provided answers. Tall trees are huge and heavy structures that need physical support and a system for lifting water from the ground and up to the leaves, where photosynthesis takes place. They also need to transport the sugars and other useful substances made in the leaves back down to the rest of the plant.

The migration of plants on to land in the Silurian – and their subsequent diversification – was made easier by the evolution of multifunctional conducting tissues called vascular bundles. These clusters of microscopic tubes develop from cylindrical cells placed end to end.

One set of tubes, the xylem vessels, are

A tree trunk is massive scaffolding for the living foliage. Its internal structure of layers of vessels gives it immense strength but allows flexibility when it is battered by strong winds.

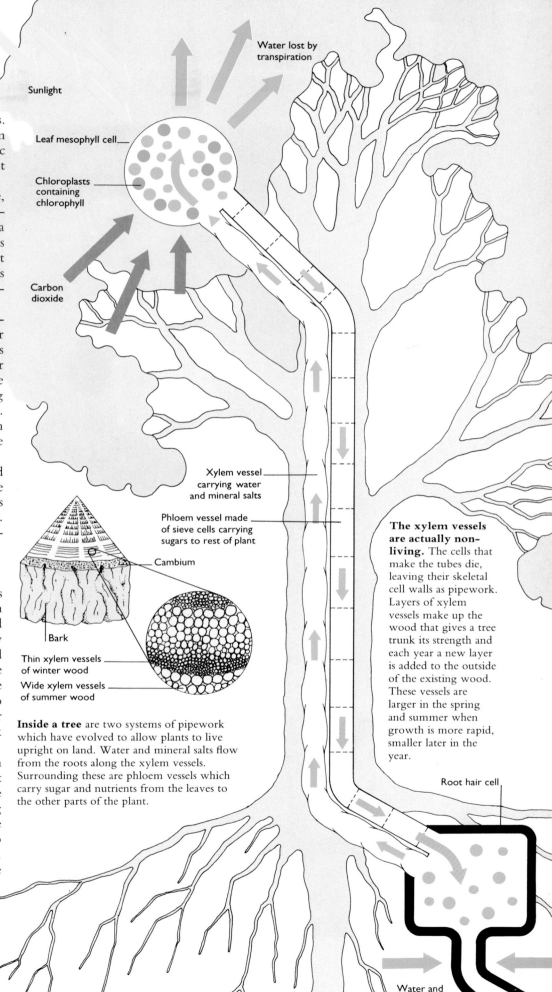

Sunlight

Water lost by transpiration

Leaf mesophyll cell

Chloroplasts containing chlorophyll

Carbon dioxide

Xylem vessel carrying water and mineral salts

Phloem vessel made of sieve cells carrying sugars to rest of plant

Cambium

Bark

Thin xylem vessels of winter wood

Wide xylem vessels of summer wood

Inside a tree are two systems of pipework which have evolved to allow plants to live upright on land. Water and mineral salts flow from the roots along the xylem vessels. Surrounding these are phloem vessels which carry sugar and nutrients from the leaves to the other parts of the plant.

The xylem vessels are actually non-living. The cells that make the tubes die, leaving their skeletal cell walls as pipework. Layers of xylem vessels make up the wood that gives a tree trunk its strength and each year a new layer is added to the outside of the existing wood. These vessels are larger in the spring and summer when growth is more rapid, smaller later in the year.

Root hair cell

Water and mineral salts enter root hair

The giant panda has a sixth digit – a "thumb" which has evolved from the radial sesamoid wristbone. This adaptation helps the panda grip bamboo stems more efficiently. Neither bears nor raccoons – the two groups most closely related to the giant panda – possess such an adaptation.

Bear

Giant panda

Radial sesamoid "thumb"

Raccoon

wide-bore ducts made of tough and flexible cellulose, further hardened in trees and other woody plants with materials such as lignin and hemicellulose. They carry water and mineral salts from the roots to the rest of the plant. When aggregated together, they generate strength. Trees grow thicker each year by adding a circumferential layer of xylem vessels – an annual ring of wood.

The bulk of any tree is a non-living scaffolding of parallel xylem tubes that not only provides flexible support against the huge forces caused by lateral winds but also gives the trunk and branches the rigidity they need to stand up against the force of gravity.

HERBIVOROUS ADAPTATIONS

The sugars and other substances made in the leaves of green plants provide herbivorous animals with the basis of their nutritional strategy. Herbivores, in fact, eat all kinds of plant matter and have evolved physiological and structural adaptations to enable them to gather, digest and use vegetable foods as efficiently as possible.

In some lineages of animals, in which plant-eating has a long unbroken history, a wide range of structural and metabolic features has been adapted for using plant foods. The teeth, jaw structure and hugely modified gut organization of bovids, such as cows, point to a long history of grass consumption. Grass and the leaves of other types of flowering plants are not intrinsically nutrient-rich for animals with a normal digestive physiology. Most of the bulk of such material consists of water and indigestible cellulose.

Cows and some other specialized herbivores, such as termites, have circumvented this nutritional paucity in cellulose-rich foods by developing a digestive physiology and gut structure based on fermentation of the plant food by microorganisms. This fermentation breaks down the cellulose to nutrients that the cow or termite can use.

THE GIANT PANDA

The evolutionary history of the giant panda provides a more vivid insight into

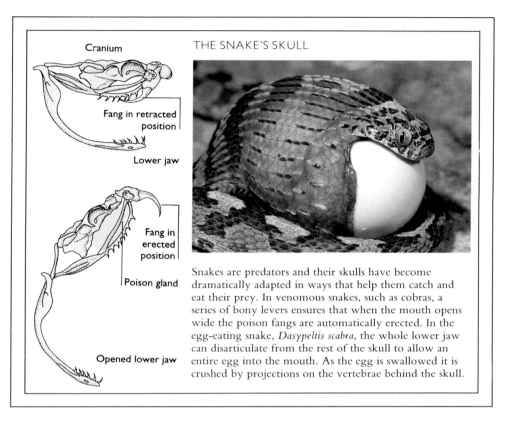

THE SNAKE'S SKULL

Cranium

Fang in retracted position

Lower jaw

Fang in erected position

Poison gland

Opened lower jaw

Snakes are predators and their skulls have become dramatically adapted in ways that help them catch and eat their prey. In venomous snakes, such as cobras, a series of bony levers ensures that when the mouth opens wide the poison fangs are automatically erected. In the egg-eating snake, *Dasypeltis scabra*, the whole lower jaw can disarticulate from the rest of the skull to allow an entire egg into the mouth. As the egg is swallowed it is crushed by projections on the vertebrae behind the skull.

the workings of adaptational forces. This bearlike creature has become a herbivore despite a genetic starting point that predisposed it to being carnivorous and now relies almost solely on a diet of bamboo leaves, stalks and shoots.

The giant panda, *Ailuropoda melanoleuca*, has been an enigma since the French missionary and naturalist Père Armand David first described it on a visit to China in 1869. Even today there is controversy about whether the giant panda and the smaller red panda belong to their own family, the bear family (Ursidae) or the raccoon family (Procyonidae). Modern molecular studies have, by analysing differences in DNA and proteins, estimated the genetic distances between bears, raccoons and pandas, and clarified their evolutionary relationships.

The ancestral group that gave rise to both the bears and the raccoons existed about 40 million years ago. Some 5 million years later, the precursors of the bears and the raccoons split apart. Then, after a pause of another 5 million years, the "raccoon" lineage divided into the

raccoons themselves and another group, whose only modern representative is the red panda. The giant panda belongs to a subfamily which split from the other subfamilies within the bear group about 20 million years ago.

The giant black and white panda is really a specialized bamboo-eating bear and its adaptations for this way of life are intriguing. Its tall skull encompasses huge muscles that power the jaws as broad molar teeth crush resistant, silica-strengthened bamboo branches. The panda's gut, however, has changed little from the simple, non-ruminant stomach and short intestine of its carnivore ancestors. Pandas cannot extract as many nutrients from their bamboo diet as a ruminant could. Because of this they never store enough fat to hibernate.

THE SIXTH DIGIT

Perhaps the most remarkable structural adaptation of the giant panda is the extra "thumb" on each of its forepaws. This unique sixth digit is an elongated wristbone, the radial sesamoid, which enables

the panda to grasp and strip leaves from bamboo branches with great ease.

The panda's thumb is a good example of the reality of evolution. Looking at a panda squatting for 12 hours a day stripping leaves with a unique wrist digit that no mammal before or since has developed, an observer is left wondering why this animal needed to have a sixth appendage on each hand. A pre-existing digit would have been the most likely candidate for the job. But, with the other digits committed to the formation of a slashing, bearlike paw, there were obviously developmental constraints that made it difficult for one of the digits to evolve into a flexible, opposable

thumb like that of the primates. Instead, genetic variation in the wrist structure provided the evolutionary raw material for the panda's sixth digit.

ADAPTATIONS OF SNAKES
The giant panda's skull, quite different in shape from that of its nearest bear relatives, is an example of the developmental plasticity of skull bones through evolutionary change. Another more far-reaching example of this sort of structural change can be found in the skulls of snakes. Here, the skull adaptations are part of a suite of skeletal alterations that adapted the snakes for an underground, burrowing, carnivorous lifestyle.

Snakes evolved from a group of limbed lizards that had probably adopted an underground burrowing way of life. The first known snake fossils, such as *Palaeophis*, come from the late Cretaceous. Compared with these, all modern snakes have greatly modified skulls (see box p. 97). In species that kill with poison fangs, slender bones form a series of levers that erect a huge poison-delivering fang on each side of the upper jaw.

Without limbs, the snakes have to move by making their bodies undulate. The vertebrae can articulate flexibly with one another to allow these waves of motion. Pointed scales, each of which is supported by an internal rib, line the

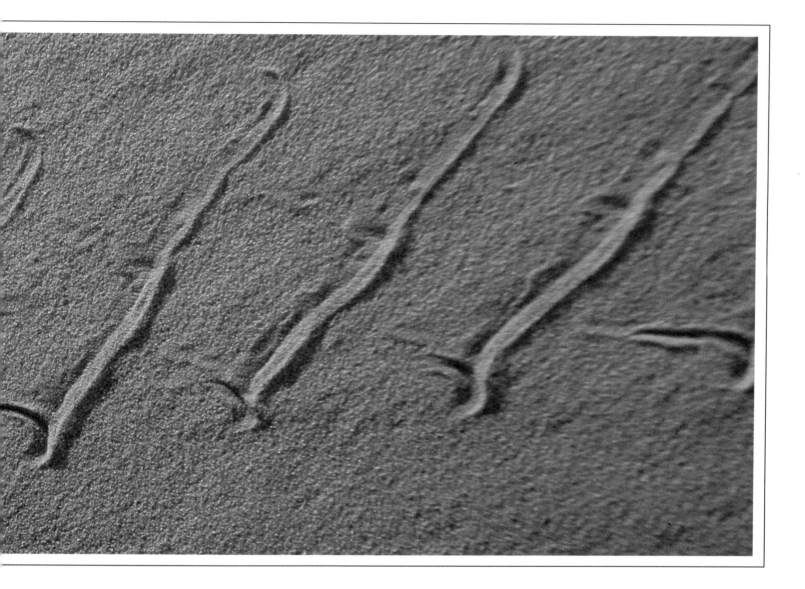

underside of the body and permit additional purchase on tunnel walls and on a variety of surfaces.

MARINE ENVIRONMENTAL CHANGES
The evolutionary transformation of normal limbed reptiles into limbless snakes is one of the more extreme skeletal adaptations seen in land-living vertebrates. The story of whale evolution, a marine equivalent, is a tale of wholesale changes in organ systems.

Since the end of the Cretaceous 65 million years ago, small, four-legged, flesh-eating, hairy land mammals — perhaps similar to species such as *Andrewsarchus* — have been transformed into

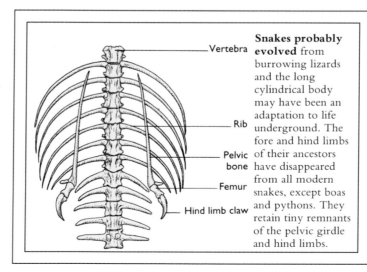

Snakes probably evolved from burrowing lizards and the long cylindrical body may have been an adaptation to life underground. The fore and hind limbs of their ancestors have disappeared from all modern snakes, except boas and pythons. They retain tiny remnants of the pelvic girdle and hind limbs.

Vertebra
Rib
Pelvic bone
Femur
Hind limb claw

Most snakes move by side-to-side undulations of the body, pushing against irregularities on the ground. This desert sidewinder, *Bitis peringueyi*, has evolved a slightly different way of moving, suited to soft sand. It throws its body into lateral waves, only two short sections of which touch the ground. The snake pushes itself sideways from these two points, leaving a trail of parallel markings.

99

enormous sea-going creatures, each with a swimming tail, flippers and a sleek, hairless, streamlined shape.

This transformation was probably stimulated by the competitive change in the marine environment, wrought by the mass extinctions at the end of the Cretaceous (pp. 186–87). Until this time the largest predators in the seas were large sharks and huge reptiles. Forms such as plesiosaurs and pliosaurs (with four large paddles resembling aerofoils) and ichthyosaurs (shaped much like a modern dolphin with fore and rear flippers, a dorsal fin and tail flukes) filled most of the large predator niches in the sea. None of the massive reptilian carnivores survived into Tertiary times. The absence of major competitors may have enabled land-based carnivorous mammals, already able to make temporary marine hunting excursions, to adapt further for a marine existence.

EARLY WHALES – "KING LIZARD"

According to the fossil record, whale ancestry harks back some 54 million years to the archaeocetes – four-limbed and seal-like, these creatures were not completely specialized for life in water. Their ear bones, for instance, did not show the characteristic changes for underwater functioning seen in more advanced forms. Indeed, the fossils were often found with those of land animals, indicating perhaps that they were only temporary marine residents.

By the middle of the Eocene, these tentative forays in constructing sea-dwelling carnivorous mammals had developed into such enormous elongate species as *Basilosaurus isis*, whose remarkable 50-ft (15-m) fossils have been dug out of marine sandstones in the deserts of north-central Egypt. *Basilosaurus*, discovered early in the 19th century, was erroneously imagined to be a giant reptile, hence its generic name, which means "king lizard". In 1839, the British paleontologist Richard Owen realized from skeletal details that this mighty creature was not only a mammal but a whalelike one.

In 1990, University of Michigan paleontologists revealed the mosaic of

The ancestors of the whales may have resembled the long-extinct *Andrewsarchus*.

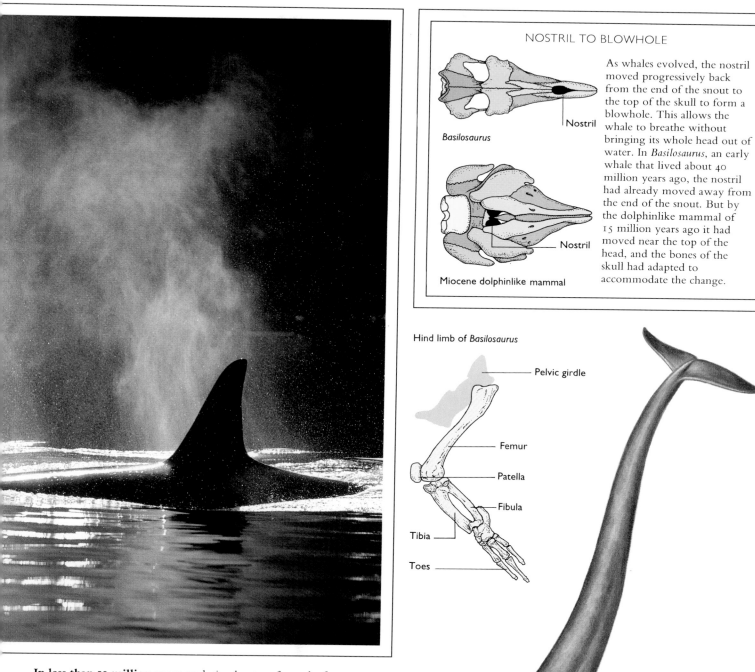

Basilosaurus

Nostril

Nostril

Miocene dolphinlike mammal

As whales evolved, the nostril moved progressively back from the end of the snout to the top of the skull to form a blowhole. This allows the whale to breathe without bringing its whole head out of water. In *Basilosaurus*, an early whale that lived about 40 million years ago, the nostril had already moved away from the end of the snout. But by the dolphinlike mammal of 15 million years ago it had moved near the top of the head, and the bones of the skull had adapted to accommodate the change.

Hind limb of *Basilosaurus*

Pelvic girdle

Femur

Patella

Fibula

Tibia

Toes

In less than 50 million years evolution has transformed a four-legged, land mammal into a sleek sea creature. Killer whales, *Orcinus orca*, (*above*) are among the fastest moving and most predatory of all modern whales.

Like a modern whale, *Basilosaurus* had a streamlined body with tail flukes and front flippers. But unlike modern whales, which have no trace of their ancestors' hind limbs, it had small back legs which could be flexed and extended. They may have been used as graspers for underwater mating.

traits exhibited by *Basilosaurus* and placed the animal halfway between modern whales and the terrestrial antecedents of the group. They already knew about its large carnivorous teeth and how its nostrils had moved back from the tip of the snout toward the top of the head.

As well as possessing large fore flippers *Basilosaurus* had hind limbs with feet that were far too small to have been used as propulsive paddles. Movement patterns suggested by the bone articulations in the limb fossils have suggested that they were claspers with which the early whales held each other while mating in the water.

BODY PLAN CHANGES

Ancient whales of Eocene times such as *Basilosaurus* probably gave rise, by adaptive routes which are not well docu-mented in the fossil record, to the two modern groupings of cetaceans. These are the toothed whales (Odontoceti) and the whalebone, or baleen, whales (Mysticeti). Modern Odontoceti, whose jaws carry ordinary carnivorous teeth, include the specialized river dolphins, killer whales, marine dolphins and the sperm whale. The Mysticeti are filter-feeding species that lose their functional teeth as adults.

In all these forms, the specialized structural adaptations required for an aquatic existence have brought about profound changes to the typical mammalian body plan. Together, the adaptations are probably more extreme than those which have occurred in any other mammal group of comparable size.

The design constraints imposed by movement through a fluid have resulted in a remarkable convergence between the evolutionary direction of whale adaptations and those seen in fish and the earlier marine plesiosaurs. All these groups have converged on a smooth streamlined outline offering minimum water resistance.

Modern whales have retained the fore limbs as flippers for steering and some propulsive activity. The main thrust for swimming comes from the up-and-down movements of the horizontal tail flukes, resulting in sinusoidal movements of the whole body. The flukes are exactly the same in general form and function as the vertical tail fins of fish, except they have been displaced through 90 degrees.

Fast-swimming cetaceans, such as killer whales, dolphins and the porpoise, have a triangular dorsal fin extraordinar-ily like that of a shark, which acts as a

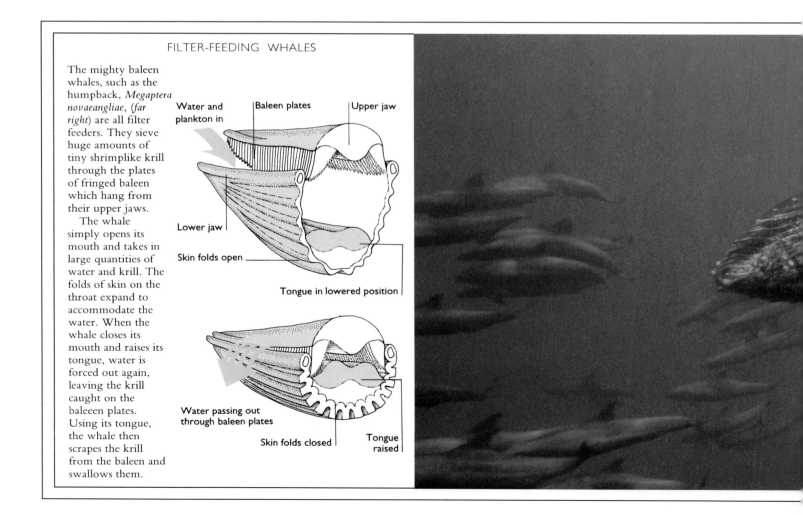

FILTER-FEEDING WHALES

The mighty baleen whales, such as the humpback, *Megaptera novaeangliae*, *(far right)* are all filter feeders. They sieve huge amounts of tiny shrimplike krill through the plates of fringed baleen which hang from their upper jaws.

The whale simply opens its mouth and takes in large quantities of water and krill. The folds of skin on the throat expand to accommodate the water. When the whale closes its mouth and raises its tongue, water is forced out again, leaving the krill caught on the baleen plates. Using its tongue, the whale then scrapes the krill from the baleen and swallows them.

Water and plankton in | Baleen plates | Upper jaw

Lower jaw

Skin folds open

Tongue in lowered position

Water passing out through baleen plates

Skin folds closed

Tongue raised

directional stabilizer in fast motion. All in all, these modifications have changed the terrestrial mammalian "building material" of the early Tertiary into a finely tuned aquatic machine. Something akin to a dog has been transformed by structural adaptations into a giant, air-breathing "fish".

The hind limbs have been lost and may have been turned into internal supports for reproductive structures. Hair has disappeared from the skin to leave a smooth, blubber-insulated, low-drag surface for swimming. Tail flukes and dorsal fins have arisen anew. The air-breathing respiration of their terrestrial ancestors has been retained but modified. Special valves have evolved for shutting the nostrils – or blowhole – while diving, and the nostrils themselves have moved to the top of the head.

FEEDING ADAPTATIONS

Specialized nutritional "tactics" form part of the whales' evolutionary diversification. Many of the toothed species, such as dolphins, are essentially fish eaters with many simple peg-shaped teeth for grasping their prey. In the killer whale this basic hunting technique has been scaled up so that the powerful toothed jaws can disable and swallow prey as large as seals and small cetaceans. In some of the toothed whales that specialize in eating soft-bodied squids, the teeth are reduced – to a solitary lower pair in the case of the bottle-nosed whale.

The baleen whales – the largest animals that have ever lived – filter feed on tiny marine invertebrates, or plankton. The largest of all is the blue whale, *Balaenoptera musculus*, which can grow to 100 ft (30 m) long and weigh 143 tons (130 tonnes), as much as 20 elephants. It feeds on shrimplike krill in polar seas.

Like all baleen whales, the blue whale has highly adapted jaws. Teeth in the adult mammal are completely absent – they make only a transitory appearance in the fetal whale. The plankton food is strained out of the water through the plates of baleen hanging from the upper jaw (see box).

Putative chordate ancestors of the vertebrates – the sea squirts (pp. 88–89) – were tiny marine filter feeders, straining marine food with a perforated pharynx. The blue whale, at the tip of one modern branch of the evolutionary tree of vertebrates, has a bulk billions of times that of a sea squirt yet has returned to the feeding strategy of its ancestors. Like the minute sea squirt, it filter feeds but on huge mouthfuls of crustaceans.

ADAPTATIONS OF PHYSIOLOGY

The green camouflage of a leaf insect, the long spines of a cactus and the sixth digit of a giant panda are all highly visible adaptations that confer selective advantage in particular environments. The leaf insect can hide from its predators amid green leaves, the spiny cactus gains protection from foraging desert herbivores and the panda strips leaves from bamboo plants with ease.

Adaptations of equal significance manifest themselves inside an organism where, often at the molecular level, they shape its physiology and biochemistry. Beneficial mutations in genes generate new proteins that modify or control such internal processes as water loss, respiration, digestion and photosynthesis. The resulting physiological and biochemical adaptations tune an organism's metabolic machinery to a particular set of environmental conditions.

The highly specialized animals, plants and microbes which survive in extreme conditions such as tropical heat, desert drought or high altitudes provide startling proof of the power of adaptation to mould the internal workings of living things.

The haemoglobin of an animal living in low levels of oxygen at high altitudes is a clear example of how a gene mutation, by causing a single amino acid substitution in a protein, produces an adaptive biochemical change. Haemoglobin, a protein which contains iron, is responsible for carrying oxygen in the blood. In response to the selection pressure of reduced levels of oxygen in the air, the haemoglobin molecules of high-altitude animals have been modified to make oxygen bind to them more strongly. This has been achieved by substituting a single amino acid in the haemoglobin molecule. The South American llama possesses such an adaptation, while its lowland cousin, the camel, does not.

The darkling beetle of southwestern Africa displays a typical physiological adaptation to the hot, dry conditions of tropical deserts, where controlling water loss and obtaining water are vital. When fog rolls in from the Atlantic Ocean the beetle moves to the crest of a sand dune in the Namib Desert and stands on its head with its wing cases and back legs facing the breeze. Water condensing on its wax-waterproofed cuticle runs down the beetle's tilted body and into its mouth.

Standing on its head, this darkling beetle, *Onymacris unguicularis,* collects every drop of condensing moisture that rolls down its body. In one such drinking session it can increase its body weight by 40 percent.

Adaptations to minimize water loss help the *Welwitschia* plant survive in even the driest parts of the Namib Desert. The plant opens its stomatal pores at night, when the air is cooler, to take in carbon dioxide. It then stores this for photosynthesis in daylight hours.

Adaptations for resisting desiccation are a feature of desert plants all over the world. To reduce water loss through evaporation, cacti and succulents have thick waxy cuticles and their stomatal pores, which permit carbon dioxide to enter and water vapour to escape, are sheltered from the heat of the air. These plants also have large internal water stores. Several shrubs from the Namib Desert consist solely of green photosynthesizing stems with stomatal pores hidden in deep grooves in the stem sides. Immediately after a rare rain storm a few short-lived leaves appear, boosting photosynthesis during the brief period of plentiful water.

The desert-adapted *Welwitschia* of the Namib, a gymnosperm classified with the conifers but resembling a flat-leaved fern, may survive for over a thousand years with only two constantly growing leaves. *Welwitschia* is thought to be the only gymnosperm that is able to use a method of photosynthesis known as CAM (p. 109) which reduces water loss.

Desert-dwelling herbivorous mammals, such as the eland and the oryx from Africa, possess a remarkable tolerance of drought conditions. Both these animals can live in intense heat without ever drinking at a water source. They reduce their water loss by producing dry fecal droppings and concentrated urine. The only significant water loss — still an extremely small amount — takes place by evaporation from the skin. This accounts for 60 percent of water lost from the oryx.

Oryxes satisfy their water needs through diet and from the metabolism of sugars, which releases water as a by-product. The grasses they eat are only about 1 percent water in the daytime heat. But after sunset, as the relative humidity of the air increases, dew forms on the grass. At night, when the oryxes feed, the dew-soaked grass blades can be as much as 42 percent water.

When air temperatures are high most mammals maintain their body temperature at about 98°F (37°C) by losing water as sweat from their skin. The water that evaporates carries away much heat. Since the oryx cannot use unlimited skin evaporation to cool itself, it side-steps the need for temperature control. Like the camel, the oryx lets its daytime temperature rise by as much as 11–13°F (6–7°C) by switching off sweat secretion. At night, the animal regains a normal body

Brain

Nasal passages

Veins from nasal passages

Artery to brain

Heat exchange blood vessels

Cold

Cooled blood to brain

Cold blood from nose

Heat exchange area

Artery

Hot

Vein

A counterflow heat-exchange system has evolved in some desert creatures, such as the oryx, to enable them to function in extreme heat. In the cavernous sinus beneath the brain hot arterial blood loses its heat to cooler venous blood. The cooled arterial blood then flows to the brain, so stopping this vital organ from overheating.

temperature when it loses heat through conduction and radiation, neither of which uses up water.

NEW ROUTES FOR PHOTOSYNTHESIS

The pivotal metabolic machinery of green plants is geared to promoting photosynthesis, a process which uses sunlight to manufacture carbohydrates from carbon dioxide and water. Hot, dry conditions force plants to contend with two problems. First, the loss of water

through their pores by evaporation. Carbon dioxide enters the pores, moving to chloroplasts in leaf cells where photosynthesis occurs. But as carbon dioxide enters the leaf cells, water vapour escapes and must be replaced.

The second problem is that oxygen – a product of photosynthesis – inhibits an early step in the fixing of carbon dioxide to form sugars. Moreover, this oxygen inhibition increases as temperatures rise. The step affected involves the formation

of phosphoglyceric acid, a chemical compound that contains three carbon atoms. Because of this vital step normal photosynthesis has been given the title "C_3 photosynthesis".

In temperate climates, most plants – from oak trees to dandelions – use C_3 photosynthesis because water losses can be easily replaced by taking up water from the soil and temperatures are never high enough to encourage oxygen inhibition. However, in constantly hot, dry

conditions where water loss and oxygen inhibition bedevil C3 photosynthesis, a variety of plant species have found alternative metabolic routes which have clear selectional advantages.

Cacti and succulents, for example, employ Crassulacean Acid Metabolism (CAM), named after the Crassulaceae family of stonecrop succulents in which it was first described. These plants close their pores by day to avoid excess evaporation and open them in the cool of the night. Carbon dioxide entering the plants is fixed in the four-carbon (C4) compound oxaloacetic acid, which acts as a temporary store of carbon dioxide during the hours of darkness. As the sun rises the pores close and the carbon dioxide is released from its C4 store for use in photosynthesis.

In the 1960s, scientists discovered that both phases of CAM photosynthesis happen simultaneously in various highly productive, fast-growing tropical plants, such as maize, sugar cane and papyrus.

Tropical grasses, such as sugar cane and papyrus, *Cyperus papyrus* (*left*), make use of a specialized form of photosynthesis known as C4. This adaptation helps the plants grow fast with minimum water loss in hot, dry conditions.

The temporary fixation of carbon dioxide in a C4 compound continues around the clock in normal leaf mesophyll cells. During the day, the C4 store is transferred to sheaths of chloroplast-rich cells surrounding the vascular bundles in the leaves. Here, the so-called C4 photosynthesis which occurs seems to be unaffected by oxygen inhibition and is highly efficient in hot, bright conditions.

The selectional advantages of C4 photosynthesis are climatically tuned. Thus, only a few C4 grasses grow in the cool climate of the British Isles, whereas in sunny California some 80 percent of grass species are C4 photosynthesizers.

ADAPTATIONS FOR DIVING

Over the last 50 million years or so, a range of land mammals – the ancestors of whales, seals, sea lions and walruses, for example – made the transition to life in the oceans. Seals exhibit the kinds of adaptations necessary for an animal, whose body is geared to living in air at normal pressure, to dive 1,650 ft (500 m) under the sea and to remain there at much greater pressure for more than an hour. Adaptations to the workings of the blood system, respiration and cell metabolism work together to enable these deep

diving seals to operate efficiently in extreme conditions that would kill a human in seconds.

The Weddell seal, for instance, dives to great depths in its search for Antarctic cod in the cold waters off Antarctica. A thick layer of insulating blubber keeps out the intense cold of the water, while an enormous volume of blood, extraordinarily rich in haemoglobin, gives phenomenal oxygen-carrying power.

As a seal starts to dive, its spleen releases up to 5.3 gallons (24 litres) of red cells to allow its blood to carry extra oxygen. A dramatic fall in heart rate accompanies a restriction of the blood supply to only the essential organs. At depths of 1,650 ft (500 m) the seal's lungs collapse completely and residual air is stored in spaces in the strengthened trachea. Without this strategy, which prevents an excess of nitrogen from dissolving in the blood, the seal would suffer from "the bends" (the dangerous release of bubbles of nitrogen in the bloodstream) when it surfaces. During the dive non-essential tissues work anaerobically, producing waste products such as lactic acid, which have to be broken down when the seal breathes oxygen once more.

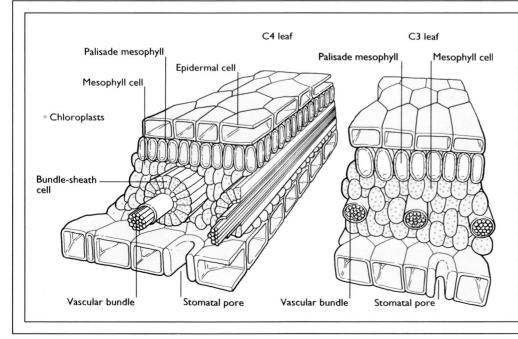

Palisade mesophyll
C4 leaf
C3 leaf
Palisade mesophyll
Mesophyll cell
Epidermal cell
Mesophyll cell
Chloroplasts
Bundle-sheath cell
Vascular bundle
Stomatal pore
Vascular bundle
Stomatal pore

C3 AND C4 PLANTS

The internal structure of most leaves is organized for normal (C3) photosynthesis. But some plants are specialized for a variant of photosynthesis known as C4. In a normal leaf chloroplasts are contained in mesophyll cells scattered throughout the thickness of the leaf. In a C4 leaf an extra set of chloroplast-containing mesophyll cells are clustered in sheaths around the leaf's vascular bundles.

The names C3 and C4 given to these two sorts of plant refer to the number of carbon atoms in the molecule in which carbon dioxide is first trapped when it enters through the stomatal pores. In the C4 leaf, carbon dioxide is trapped in a C4 compound in ordinary mesophyll cells before being transferred to the sheath cells, where it can be used in normal C3 photosynthesis.

DIVING ADAPTATIONS

One of the deepest diving of all seals is the Weddell seal, *Leptonychotes weddelli* (*above*). Physiological adaptations help it plunge to depths of more than 1,650 ft (500 m).

Adrenal glands

Diaphragm muscle

Spinal cord

Lungs

Brain

Eye

Abdominal organs

Body wall muscle

Heart

Blubber

When a Weddell seal dives, it conserves oxygen by cutting off the supply of oxygenated blood to all but those organs essential for movement and navigation (shown in red). The heart muscle continues to receive blood but in reduced quantities.

LIFE IN UTTER DARKNESS

In 1977, the research submersible *Alvin* explored the Pacific Ocean to the northeast of the Galápagos Islands and discovered apparently impossible densities of life in the utter darkness of the ocean floor. These flourishing communities were clustered around vents that spewed out hot, mineral-rich water from the rocks some $1\frac{1}{2}$ miles (2.5 km) below the ocean surface.

The paradox of the community, which contained clams, pogonophoran tube worms, crabs, fish and so on, was that, since plants were absent, it seemed to lack the primary producers needed to inject energy into the system. On land or in shallow sunlit water the primary producers are always plants: the products of their photosynthesis sustain all the other organisms in the community.

On what, then, do the animals of the deep-sea vent feed? Research has revealed that bacteria in this and other similar communities (which have since been found all over the world) can generate energy from the hydrogen sulphide in the mineral-rich water. Bacteria, not plants, are therefore the primary producers of the deep-sea vent ecosystem. Some community members feed directly on the bacteria. Many others, such as the pogonophorans and the clams, have taken the bacteria into their bodies where, through symbiosis, they gain direct nutritional benefit from the microbes' metabolic alchemy.

ADAPTING TO COLD

The flora and fauna inhabiting the high-latitude regions of the world have evolved built-in programmes of seasonal metabolic change to cope with the recurring rigours of cold winters. These programmes are often synchronized to changing day lengths – many flowering plants, for instance, shed their leaves in response to shortening day lengths and so avoid damage by frost and ice.

Warm-blooded animals, faced with the problem of how to maintain their body temperature during the cold winter, either have to expend more energy, and therefore eat more, as temperatures fall below freezing or else

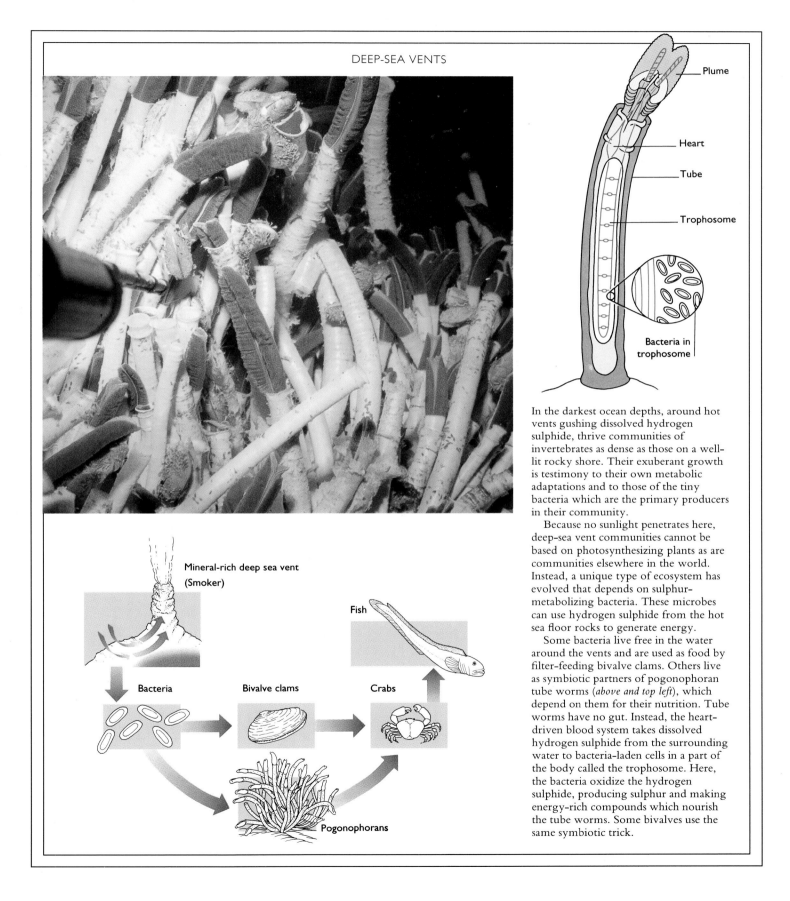

DEEP-SEA VENTS

Plume

Heart

Tube

Trophosome

Bacteria in
trophosome

Mineral-rich deep sea vent
(Smoker)

Fish

Bacteria

Bivalve clams

Crabs

Pogonophorans

In the darkest ocean depths, around hot
vents gushing dissolved hydrogen
sulphide, thrive communities of
invertebrates as dense as those on a well-
lit rocky shore. Their exuberant growth
is testimony to their own metabolic
adaptations and to those of the tiny
bacteria which are the primary producers
in their community.

Because no sunlight penetrates here,
deep-sea vent communities cannot be
based on photosynthesizing plants as are
communities elsewhere in the world.
Instead, a unique type of ecosystem has
evolved that depends on sulphur-
metabolizing bacteria. These microbes
can use hydrogen sulphide from the hot
sea floor rocks to generate energy.

Some bacteria live free in the water
around the vents and are used as food by
filter-feeding bivalve clams. Others live
as symbiotic partners of pogonophoran
tube worms (*above and top left*), which
depend on them for their nutrition. Tube
worms have no gut. Instead, the heart-
driven blood system takes dissolved
hydrogen sulphide from the surrounding
water to bacteria-laden cells in a part of
the body called the trophosome. Here,
the bacteria oxidize the hydrogen
sulphide, producing sulphur and making
energy-rich compounds which nourish
the tube worms. Some bivalves use the
same symbiotic trick.

relinquish temperature control by hibernating. Since winter food is scarce for such mammals as dormice, hedgehogs, marmots and ground squirrels, hibernation has become an evolutionary survival strategy. It is advantageous to them not to have to maintain a high body temperature through the coldest part of the year.

During hibernation, these mammals experience a decline in their metabolic activity. Their rates of breathing and heartbeat slow down and the temperature of their bodies starts to match that of the air around them. When they awake their temperature is often raised by the increased metabolism of their "brown fat" cells, a specialized fatty tissue which acts like a living electric blanket. Multitudes of mitochondria in these cells "burn" the fat, producing so much heat that the animal is warmed and its temperature returns to normal.

ADAPTATIONS FOR SPEED

Extra mitochondria also feature strongly in the achievements of the world's champion speed merchant. The pronghorn, *Antilocapra americana*, of North America can outrun every other animal in a long-distance race. Its top speed of 62 mph (100 km/h) has been bettered only by the cheetah. But whereas the cheetah

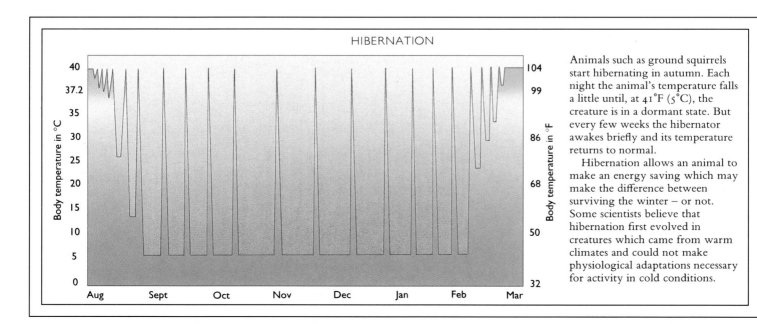

HIBERNATION

Animals such as ground squirrels start hibernating in autumn. Each night the animal's temperature falls a little until, at 41°F (5°C), the creature is in a dormant state. But every few weeks the hibernator awakes briefly and its temperature returns to normal.

Hibernation allows an animal to make an energy saving which may make the difference between surviving the winter – or not. Some scientists believe that hibernation first evolved in creatures which came from warm climates and could not make physiological adaptations necessary for activity in cold conditions.

Hibernation allows animals such as the dormouse (*left*) to escape harsh winter weather and poor food supplies. The body temperature falls and the heartbeat slows, so the minimum energy is used to keep the body going, and the animal enters a cold, sleeplike state.

can hold its speed for only a few seconds, the pronghorn can maintain speeds of 40 mph (65 km/h) for 10 minutes.

The secret of the pronghorn's ability relies on a blend of physiological and metabolic adaptations that maximize the production and expenditure of energy. A pronghorn can use oxygen three times faster than a similar-sized mammal. Its huge lungs have a vast surface area for absorbing oxygen, its blood is ultra-rich in haemoglobin and its heart can pump extremely fast. Its leg muscles are enormous and the muscle cells are crammed full of ATP-producing mitochondria.

In evolutionary terms this high-performance running ability seems to have achieved its objective – to provide a means of outstripping the wolf, the pronghorn's most important predator.

Physiological and structural adaptations help the pronghorn (*below*) use oxygen so efficiently that it holds the animal record for high-speed, long-distance running. Features such as extra-large lungs and heart help it process more oxygen than other animals of similar size.

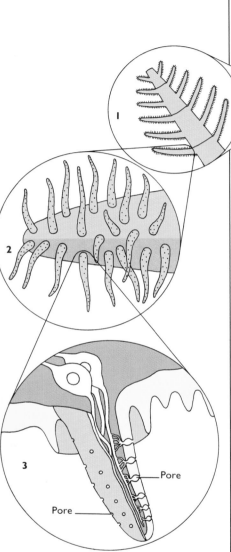

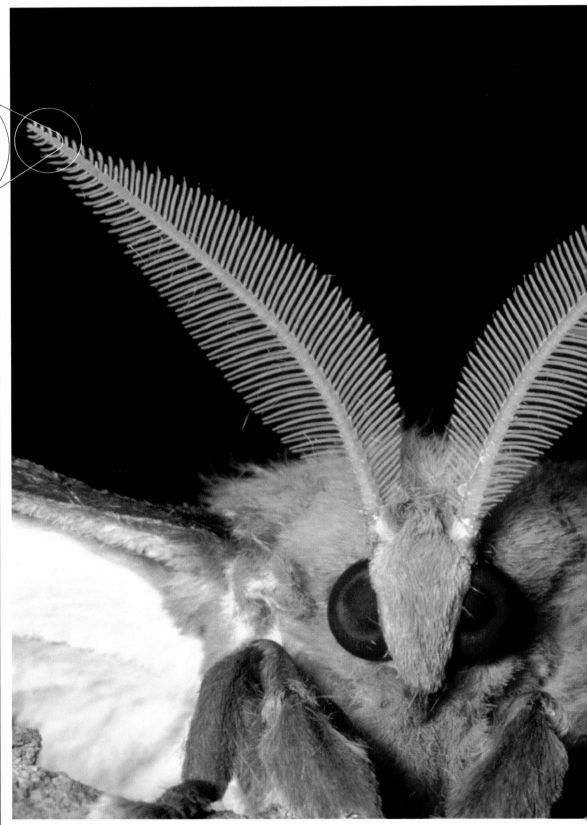

Ultrasensitive cells in the moth's antennae respond to pheromones, sexual scents, given off by females. Each side branch of the antennae (1) is itself covered in tiny branchlets which are dotted with minute pores (2). A network of fine nerves links each pore with olfactory sense cells within the branchlet (3). Scent molecules in the air enter the pores where they come into contact with nerve endings. These stimulate the nerves and relay the message to the sensory cells. The pores are "tuned" to respond only to the scent molecules of the moth's own particular species.

Pore

Pore

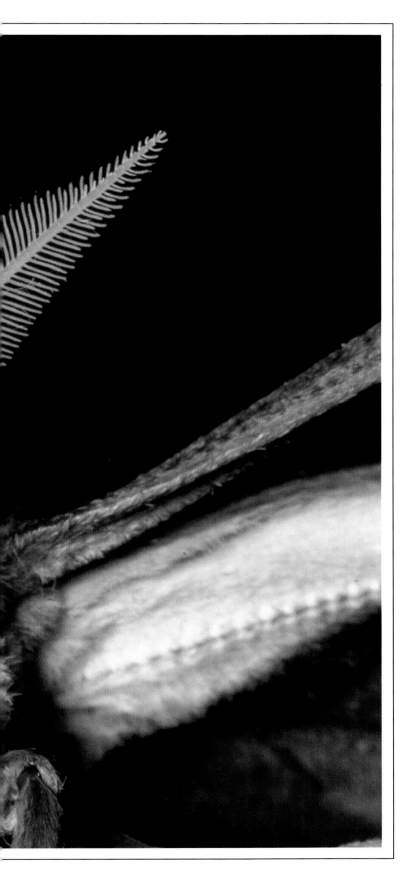

ADAPTATIONS OF SENSORY SYSTEMS

Every living thing constantly interacts with its environment. Organisms can, with varying degrees of complexity, sense changes in the world around them and readjust their behaviour or their metabolism accordingly. Even viruses have this sort of responsiveness via individual receptor molecules on their outer surface. Thus, receptor molecules on the protein coat of a virus respond to complementary molecules on the surface of its specific host cell type, enabling the virus to lock on to the cell it will subsequently invade.

A vulture, gliding on a thermal high above the baked surface of the east African savanna, spots another vulture half a mile away dropping vertically from the sky. The first vulture, seeing that the other bird has discovered a carcass on the ground, responds by gliding in the same direction to share in the spoils.

A male moth flies through the woods at night, its two hypersensitive antennae combing the air for chemical signals. They sense a few molecules of the odorous sex chemical, or pheromone, emitted by an alluring female of the same species and compel the moth to fly several hundred yards to find her.

A bacterium responds to the presence of a new sort of sugar in its watery surroundings by switching on the gene that directs the manufacture of an enzyme to digest the sugar. When the sugar disappears, the gene is switched off, thereby avoiding uneconomical production of the enzyme.

A single-celled protozoan, such as an amoeba, moves over the surface of a water plant. When it encounters food in the form of a single-celled diatom, the behaviour of its cell surface changes. At the point of contact the amoeba's membrane and cytoplasm indent, engulfing the diatom in a food vacuole, where it is digested.

These four vignettes illustrate how different organisms have evolved different adaptations for sensing and responding to their environment. In every case the organism is stimulated, on perceiving a change in its external surroundings, to alter its activity in an adaptive manner and make optimal use of the change. The vulture uses long-distance vision to locate a potential meal. And the male moth distinguishes a chemical signpost that leads him to his mate. The bacterium adapts its metabolism to take advantage of a new food source.

Using the huge sensory antennae on top of his head, the male Atlas moth, *Attacus atlas*, (*left*) can detect chemical scent signals given off by female moths. Since moths fly at night these signals help the moths to find mates.

The amoeba alters its cell membrane to ingest a prey item.

A NETWORK OF NERVES

Multicellular animals as structurally complex as jellyfish and flatworms, and those of even greater complexity, have evolved a whole network of cells – a nervous system – to handle their sensory and responsive needs. This integral system of elongated nerve cells conducts coded messages around an animal's body. The messages are composed of patterns of electrical impulses in the membranes of the nerve cells.

The basic impulse contained in a nerve message is a "spike" of electric change that is transmitted rapidly along the nerve cell. The meaning of the message depends on which type of cell sends it and on the pattern of spikes transmitted. For example, as a stimulus from the outside world grows stronger so the frequency of spikes increases.

In general, nervous systems are subdivided into different types of nerve cells, each with a specific function. Receptor nerve cells, such as the cone cells in the retina of the human eye, can sense a particular stimulus (in this case light energy) in the outside world and translate it into a coded chain of nervous impulses. Such receptor cells, wherever they are located, are sensory transducers: an external stimulus, such as light, temperature, touch or a chemical, is translated into a sequence of electrical changes in the receptor cell membrane. In complex animals these sensory transducers are built into organs, such as eyes, ears, noses or tongues. The individual structures of these sensory organs are precisely designed to channel the stimulus so that it interacts directly with the receptor cells.

Sensory nerve cells transmit coded information from the receptor cells to a coordinating centre – a nervous structure such as a brain or a ganglion – which deciphers the message and initiates a response. Effector nerve cells transmit this response back to the organs or tissues

SNAKE SENSES

Even in complete darkness the rattlesnake can find warm-blooded prey, such as this rat. The infrared receptors within the pits on the snake's head can create a "heat picture" of the immediate environment. In this, the warm rat will "glow" against the cooler surroundings.

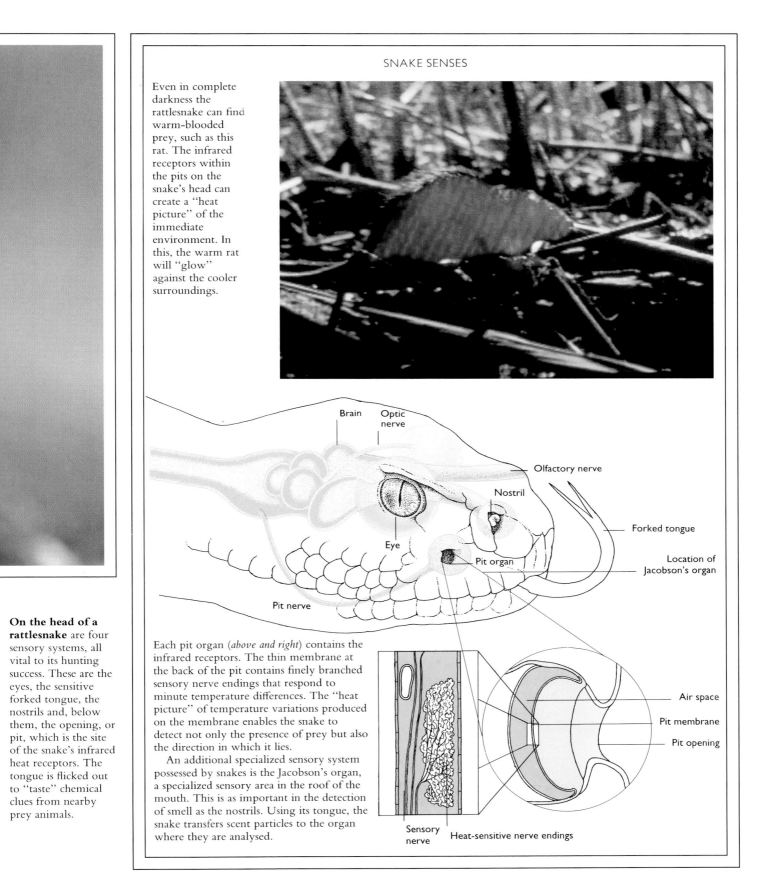

On the head of a rattlesnake are four sensory systems, all vital to its hunting success. These are the eyes, the sensitive forked tongue, the nostrils and, below them, the opening, or pit, which is the site of the snake's infrared heat receptors. The tongue is flicked out to "taste" chemical clues from nearby prey animals.

Each pit organ (*above and right*) contains the infrared receptors. The thin membrane at the back of the pit contains finely branched sensory nerve endings that respond to minute temperature differences. The "heat picture" of temperature variations produced on the membrane enables the snake to detect not only the presence of prey but also the direction in which it lies.

An additional specialized sensory system possessed by snakes is the Jacobson's organ, a specialized sensory area in the roof of the mouth. This is as important in the detection of smell as the nostrils. Using its tongue, the snake transfers scent particles to the organ where they are analysed.

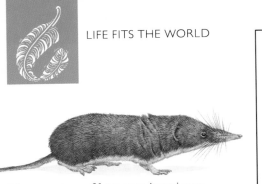

The ancestors of bats were insectivores, probably similar to this modern shrew (*above*). An intermediate stage between the land-living shrewlike creature and bats – the only mammals with true powered flight – may have structurally resembled the flying lemur (*right*) which can glide through the air on winglike flaps of skin. The earliest fossils of batlike animals date from the Eocene period, about 50 million years ago. These creatures are thought to have lived much like modern bats, flying at night and catching insects on the wing.

of the organism's body, stimulating, for example, a muscle to contract or a gland to secrete a hormone.

HEAT-SENSITIVE SNAKES
An animal with a sense organ that is acutely sensitive to a specific stimulus provides an exquisite example of a precise evolutionary adaptation. Through this adaptation the animal gains selectional advantages over its competitors, its predators or its prey. The species in two snake families, for example, have evolved novel sensory adaptations for detecting and localizing the body heat of their warm-blooded prey.

Snakes of the pit viper family, which includes rattlesnakes, the water moccasin and the copperhead, possess two heat-sensitive pits, one on each side of the head between the eye and the nostril. Members of the boa family, such as the boa constrictor and the python, have their heat-sensitive pits in rows on the scales bordering their mouth.

Each pit acts like a pin-hole camera, focusing infrared (heat) radiation on to one part of the heat-sensitive membrane within it. The snake translates the position of this "hot spot" on the membrane into a directional bearing for the source of the heat. Without visual clues, a rattlesnake can target and strike at a mouse with a maximum error of only five degrees. The sensitivity of the pit organs is astonishing – each sensory nerve fibre in the membrane can detect temperature changes as small as three

thousandths of a degree centigrade.

Sensory information on heat location is processed in the same parts of the brain as visual information, suggesting that the snake can integrate optical and infrared images into a kind of superimposed view of its surroundings. By adapting their sensors and the nervous equipment for interpreting heat clues, the cold-blooded snakes have utilized with deadly effect the high body temperatures of warm-blooded mammals and birds. In the night-time world of the infrared-sensitive rattlesnake, a mouse or nestling bird glows like a neon light. Such

Bats can fly as well as birds and catch prey in the air in complete darkness – in fact they are superbly adapted for an aerial lifestyle. Once a bat has located its prey with its radarlike sense system it can use its supple wings to scoop the prey from the air and into its mouth (*right*).

sensory equipment increases the snake's efficiency as a predator – a clear evolutionary advantage.

ULTRASONIC SENSITIVITY
Among mammalian groups bats are second only to rodents in terms of species richness. This incredible evolutionary

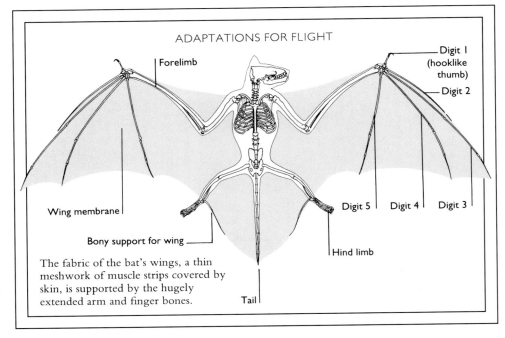

ADAPTATIONS FOR FLIGHT

Forelimb

Digit I (hooklike thumb)

Digit 2

Wing membrane

Digit 5   Digit 4   Digit 3

Bony support for wing

Hind limb

Tail

The fabric of the bat's wings, a thin meshwork of muscle strips covered by skin, is supported by the hugely extended arm and finger bones.

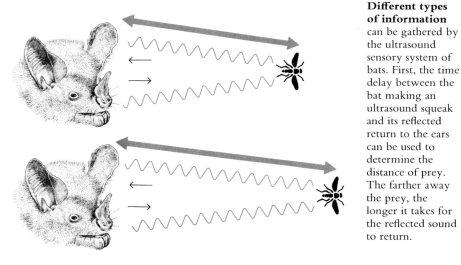

Different types of information can be gathered by the ultrasound sensory system of bats. First, the time delay between the bat making an ultrasound squeak and its reflected return to the ears can be used to determine the distance of prey. The farther away the prey, the longer it takes for the reflected sound to return.

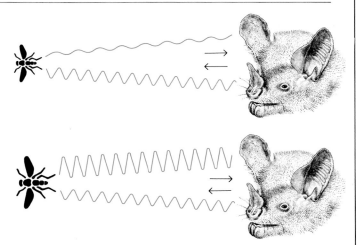

Once the distance of the prey has been computed, other information in the reflected sound can be used to determine the size of the prey. Large prey reflects back more sound than small prey – the reflected sound is louder. This volume difference is shown here by the height of the sound waves.

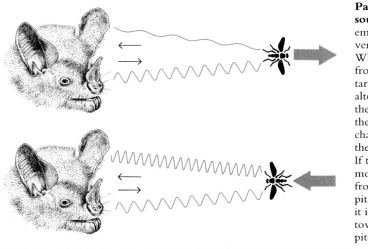

Part of the ultra-sound squeak is emitted with a very precise pitch. When reflected from a moving target this pitch alters according to the movement of the target – a change known as the Doppler shift. If the prey is moving away from the bat the pitch drops. But if it is moving toward the bat the pitch rises.

success is almost certainly the result of a major sensory innovation – ultrasonic echolocation – operating in parallel with a remarkable method of locomotion.

Bats are the only mammals that can sustain true powered flight. Their fore-arm wings, coupled with skin flaps extended between the enormously long fingers of their hands, provide sufficient lift and thrust to make them fantastically manoeuvrable in the air. Most bat species exploit their aerial skill to catch insects on the wing. But if they hunted in daylight they would be competing for food sources with innumerable species of birds. In fact, because of their sensory uniqueness, bats have become masters of the night-time skies. They can fly, locate and catch insects in complete darkness. As a result, birds and bats have, in evolutionary terms, divided the nutritional source of flying insects between them – most birds hunt by day, bats by night.

The sensory echolocation capacity of bats enables them to pinpoint insects with unerring accuracy (see diagrams) and to fly in the darkness. Bats produce ultrasonic squeaks and snorts – higher in pitch than humans can hear – that are pulses of sound. When these pulses hit an object, a pattern of echoes returns, is picked up by the bat's sensitive ears and is translated by its brain into a complex, real-time, three-dimensional map of the world around it. Bats metaphorically "see" with sound. In total darkness they can avoid the thinnest obstacle, catch the fastest flying moth and even distinguish between one insect and another.

ELECTRICAL SENSES
Bats are not the only animals to exploit sensory skills beyond the visually dominated repertoire of humans. Sharks, rays, some freshwater bony fish and even the duck-billed platypus share the ability to perceive and respond to electrical changes in their environment. There is a

The flattened nose leaf on the head of this large-eared bat, *Micronycteris megalotis*, (*far left*) is part of its sensory system. The ultrasound squeaks and chirps are "snorted" out through the nose and directed by the nose leaf into a cone in front of the bat. The huge dish-shaped ears pick up any echoes reflected back.

ELECTRICAL SENSES

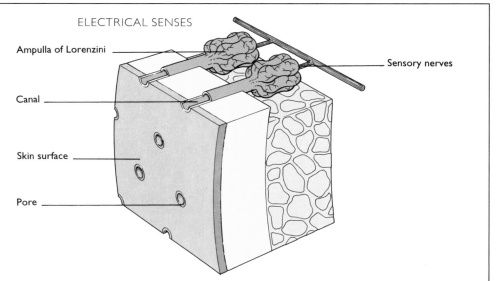

Ampulla of Lorenzini

Sensory nerves

Canal

Skin surface

Pore

The ferocious-looking head of the sand tiger shark, *Odontaspis* sp., (*below*) is dotted with fine pores. Each is the opening of an electrical receptor organ known as an ampulla of Lorenzini. These organs enable the shark to locate prey by detecting the minute electrical disturbances in the water that are set up by the muscles and nerve activity of the prey animal.

Beneath each pore is a slender canal leading down through the skin to the ampulla itself (*right*). This is a jelly-filled collection of spaces surrounded by sensory nerve endings. Electrical changes in the sea water stimulate the shark's sensory nerves.

context, of course, within which all senses may be considered electrical. Every sensory receptor cell converts an external stimulus into a chain of electrical impulses that are conveyed through the nervous system.

Animals with a true electrical sense can respond to infinitesimally small electric currents. A shark, for instance, can home in on the minute electrical disturbances generated by the nerves and muscles of prey buried in sand at the bottom of the sea.

The skin of sharks and rays bears electrical receptor cells that are connected to the nervous system. The cells form clusters, known as ampulli of Lorenzini, each of which lies at the end of a canal of electrically conducting jelly that runs inward from a pore in the skin surface. Rows of ampulli act as an extended antenna for picking up tiny electric signals. Behavioural experiments with buried or hidden prey reveal that an extremely acute chemical sense – a kind of underwater "smell" – works in tandem with the electrical receptors to help the carnivorous fish find their prey.

In 1988, electrical receptors were discovered in the flexible beak of the duck-billed platypus; they are thought to help in locating prey underwater in turbid conditions. The following year, scientists found similar receptors in the tip of the long snout of an echidna – a near relative of the platypus.

ELECTRIC SHOCKS
Some fish generate rather than sense electrical discharges. Electric eels, electric catfish and the electric ray call upon specialized muscle cells – constituting 58 percent of body weight in electric eels – to produce massive shocks that stun or kill prey, or act as a defence. The electric eel, *Electrophorus electricus*, can deliver a jolt of 600 volts over a few milliseconds with a current of half an ampere.

Until the 1960s, scientists assumed that fish generated electricity only for offensive or defensive purposes. They then discovered that certain freshwater fish, particularly some of the African mormyrids, or "elephant trunk" fish, produced electrical discharges as part of an

extremely sensitive sensory system. Their modified muscle cells emit a continuous series of 5-volt electrical discharges at the rate of about 300 per second, creating an electrical field rather like the lines of force that exist around a bar magnet.

Electrical receptors in the skin of mormyrids perceive the strength and

The platypus, *Ornithorhynchus anatinus*, possesses some surprising adaptations – it has thousands of electroreceptors in its beaklike snout. The platypus feeds at the bottom of murky lakes and rivers and the receptors help it find its prey and navigate around obstacles.

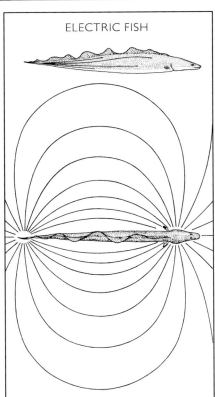

ELECTRIC FISH

Some fish have a sort of electric vision. Rows of muscle cells along the body of a mormyrid fish (*top*) generate electrical pulses that pass into the water. These set up a field of electrical changes around the fish (*above*) which is distorted by any object that enters it. Electrical receptors in the fish's skin pick up the distortions and enable the fish to locate the object.

pattern of the field. Distortions to this pattern produced by objects – animate or inanimate – are picked up by skin receptors and translated into an electrical map of the fish's immediate surroundings. This map enables the fish to navigate and find prey even in thick muddy water. Experiments reveal that the receptors can respond to voltage gradients of three hundredths of a millionth of a volt per centimetre. The ecological niche of the mormyrid fish is, like that of the bat, largely moulded by an extraordinary sensory capability.

How might such sense systems have evolved? All working muscles produce an electrical output. Presumably, mutations in the muscle cells of ancestors of electric eels and mormyrids increased that electrical activity. In the first group this was the beginning of a system for stunning prey. In the second group, such mutations produced an electrical field to which some preexisting sense cells could respond.

THE VISUAL SENSE
Human beings are highly visual creatures with excellent stereoscopic vision. Their eyes can respond to light and colour over a remarkably wide range of intensities, from near-darkness to bright tropical sunlight. The fact that the eyes of humans and most other vertebrates share a common structural ground plan suggests these sensory organs featured early in the evolution of the group. Any variations on the plan represent modifications and adaptations resulting from

the diversification of vertebrates into new habitats and niches.

The vertebrate eye is a complex living camera producing not one-off exposed negatives but a continuous, free-running, nerve-impulse translation of an optical image. In this respect the eye compares with a television camera, which transmits a moving image to a television station in the form of electronic signals.

The curved inner lining, or retina, at the back of the eye consists of millions of light-sensitive, pigmented, receptor cells called rods and cones that are wired up to the optic nerve via a complex array of interconnecting nerve fibres. The rods respond to low-intensity light of any colour, while the cones provide colour sensitivity at high light intensities.

The remaining structures of the eye ensure that an image of the outside world is focused on the retina despite wide fluctuations in light. The curved cornea and the internal lens bend light rays to form an image on the retina. Muscles change the shape of the lens to sharpen the focus on near or far objects. The eyelids and the iris control the amount of light entering the eye – iris muscles reduce the diameter of the pupil in bright light and increase it in dim light.

The eye of an octopus shares a remarkable number of basic design features with a typical vertebrate eye: a roughly spherical shape, a curved, light-sensitive retina on its rear internal surface, a transparent lens whose focus is modifiable and an iris that controls the amount of light entering the eye. This correspondence is all the more remarkable because the eyes of an octopus and a vertebrate have evolved along completely independent routes. The last common ancestor of vertebrates and octopods must have lived in the late Precambrian, long before either group possessed the sort of eyes they do today.

COMPOUND EYES

Another highly effective visual imaging system – the multifaceted compound eye – has evolved in many arthropod groups, including active visual insects, such as bees, and crustaceans, such as

lobsters. Each facet is the outer lens of a unit which, resembling a miniature telescope, points in a slightly different direction to the other facets. At the base of a unit is a tubular cluster of light-sensitive receptor cells linked to the optic nerve.

Compound eyes work in two different ways. In the first, each unit of the eye is surrounded by a sheath of dense pigment, enabling its light-sensitive cells to operate independently but to contribute to the final image pieced together by the coordinating ganglion. The unit produces the equivalent of a single pixel on a computer image. In the second, the sheathing pigment is absent, allowing the lens, mirror surfaces and light-guide properties of the units to produce a single consolidated image at the level of the

light receptor cells. This so-called superposition method may be especially efficient in conditions of low light intensity.

The varied evolutionary raw material offered by arthropod, mollusc and vertebrate lines has developed only two basic operating systems for high-definition visual sensitivity. Octopods and vertebrates have developed independent variants of one system, arthropods the other. The immensely long evolutionary lineage of these vital receptor systems is manifested in the beautifully faceted compound eyes of the 500 million-year-old trilobites. These ancient and long-extinct arthropods employed the same mechanism to form optical images of the sea bottom as honeybees do today when they home in on a flower head.

THE SENSORY OCTOPUS

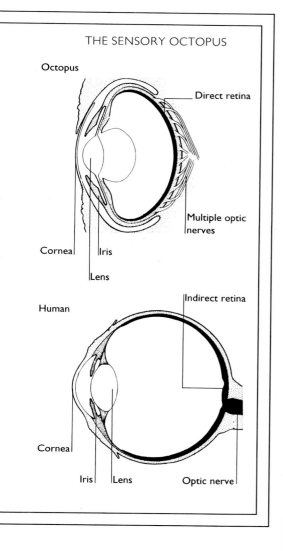

Octopus

Direct retina

Multiple optic nerves

Cornea Iris

Lens

Human

Indirect retina

Cornea

Iris Lens

Optic nerve

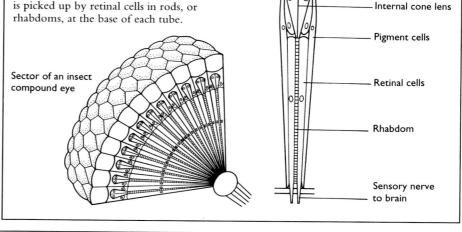

The compound eyes of insects are made up of hundreds of tubular elements, each with a facet at its outer face. The surface of the eye is thus a regular grid of these tiny facets, visible in this magnified image of a deer fly eye (*above*). Light passing through the facets is picked up by retinal cells in rods, or rhabdoms, at the base of each tube.

Sector of an insect compound eye

One unit of an insect compound eye

Surface lens facet

Internal cone lens

Pigment cells

Retinal cells

Rhabdom

Sensory nerve to brain

Highly sensitive tactile and visual abilities have evolved in the octopus (*above left*). The many suckers on each of its eight arms are mechanical devices for grasping objects and helping movement but they also contain many sensory nerves. These enable the suckers to sense the size, shape and "taste" of the object they touch.

The eyes of the octopus are probably the most sophisticated of any invertebrate, built on much the same principles as mammalian eyes. Despite their similarity, the two systems have evolved separately – two lines of development have produced much the same answer to the problems of high-resolution vision. The evolutionary gulf between the two is revealed by a comparison of the retinas, which are built on completely different principles. The octopus has a direct retina in which the photoreceptor cells are nearest the incoming light. The human eye has an indirect retina – the photoreceptors point away from the light and are covered by a layer of other nerve cells and blood vessels.

ADAPTATIONS OF BEHAVIOUR

Mighty stags locking their antlers in combat, exotic birds performing a dainty courtship dance, lively young otters playing – almost anything an animal does is a form of behaviour. And, just like any other inheritable characteristic, behaviour is subject to evolutionary adaptation, since animal species respond to changes in their environment by altering their activities, by adapting their behaviour to suit the new conditions.

The concept of behaviour is usually associated with animals that have brains, nervous systems and muscles which control and coordinate movements, but even green plants, fungi and microorganisms "behave" in their more restricted ways. Plants have tropisms – movements of parts of the plant toward or away from an important stimulus, such as light, water or gravity. Microorganisms stream toward a food source dissolved in the water they live in. Some bacteria even contain tiny magnetic particles which enable them to swim in a particular direction along the Earth's magnetic field.

Patterns of behaviour are inheritable, but the extent to which they are predetermined by genetic constitution is still the subject of controversy. Some behaviour traits, such as web building by spiders, are instinctive and automatic, directly passed down from generation to generation. But others, in those organisms such as mammals, with a genetically determined capacity to modify behaviour from past experience, are learned, so the inheritable basis is partly indirect.

Most experts agree that the two sorts of behaviour coexist, but that the balance between them differs in different groups of animals. The behaviour of most insects, for instance, is highly stereotyped and instinctive, while most primates show strong evidence of modifiable learned behaviour. Both strategies are efficient for the organisms concerned.

Adaptations of behaviour are important evolutionary traits, just as much as different wing lengths in birds or varieties of haemoglobin molecules in humans. And like physical features, behaviour traits can dramatically affect the survival and reproductive success of individual creatures.

American scientist E. O. Wilson, the father of modern sociobiology, has pointed out that an animal's behaviour can respond more rapidly and sensitively to changes in the environment than can other attributes such as body structure.

In the breeding season male ruffs, *Philomachus pugnax*, perform leks, or daily communal displays designed to attract mates. Weaker birds drop out of the contest and dominant males emerge which mate with most of the females.

RITUAL STRUGGLES

Male Siamese fighting fish, *Betta splendens*, defend their breeding territories in battles that are essentially ritualized threats – they do not generally hurt one another. Contestants display their fins, turn away briefly, then resume the aggressive posture until one backs down. The fish do not have to learn this behaviour – it is inborn.

1

2

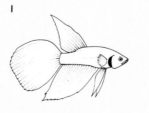

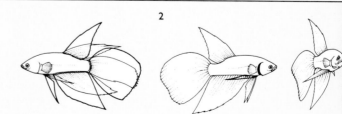

When selection pressures, such as climate or the type of food available, alter, the easiest adjustment for an organism to make is in its behaviour. Physical adaptations would usually take place much more slowly.

Just as an internal organ can be dissected, an organism's behavioural repertoire can be broken down into components which can then be separately identified, studied and compared with others. In the same way that anatomical features of species can reveal a common evolutionary ancestor, behavioural traits can be shared between different organisms with a common ancestral species.

Analysis of stereotyped behavioural traits linked with breeding and communication in a group of species reveals how the degree of physical relatedness squares with behaviour traits. Species that are closely related structurally often share most behaviour patterns, while distantly related groups share far fewer.

The birds in the order Pelicaniformes, which includes cormorants, pelicans, gannets, boobies, darters, frigatebirds and tropicbirds, have been studied in this way. Of eight key behavioural traits,

Each spring, male red deer, *Cervus elaphus*, lock antlers and engage in ruts, fierce battles for the ownership of harems of females. Only those deer that are big and strong enough to win such battles finally achieve success and breed. Such behavioural systems ensure that the males that do breed are passing on genes associated with strength and stamina. The "test" of the rut means that males without such characteristics are unlikely to be able to pass on their genes to the next generation.

such as pre-landing calls, bowing and presenting nest material to a mate, the closely related gannets and boobies shared seven. But these two groups shared only two traits with their more distant relatives, the tropicbirds.

AGGRESSIVE BEHAVIOUR

One obvious and familiar form of behaviour is aggression, which includes threats, submissive postures and chases as well as true physical conflict.

An animal may often resort to aggression in order to defend its territory or valuable resources. Male songbirds noisily protect their nest sites; honeybees attack intruders near a hive. Aggressive behaviour is also used to maintain dominance, a higher position in the "pecking order", by many social animals such as monkeys. For the spotted hyena, for example, extreme aggression is such an important part of its hunting behaviour that the young are snapping and snarling at one another as soon as they are born.

Some of the most dramatic forms of ritualized aggression are enacted by male animals in their struggles for dominance and mates. Male seals and deer battle to become the sole mate of a "harem" of females. A huge selection pressure therefore exists for those males to become super-efficient fighters – the best fighters get to mate with the most females. This sexual selection has driven evolutionary change to produce large, strong, aggressive males in these species. And some, such as the antler-bearing deer, have also become structurally adapted, with special body parts serving as weapons for conflict.

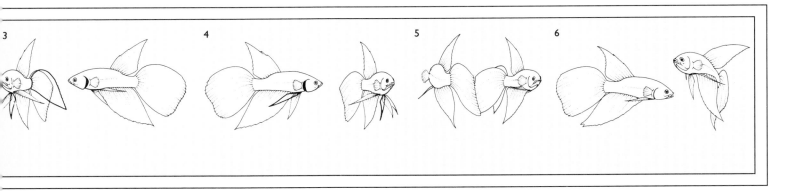

THE BEGINNINGS OF BEHAVIOUR

After the fertilization of an egg by a sperm, an embryo forms which grows into a fetal animal. This then hatches or is born into the world as an independent creature. During that period of growth the unfolding developmental pro-grammes dictated by the animal's genes define the form of its body. Part of that gene-defined process builds the animal's brain, sense organs, nervous system and muscles – the equipment it needs in order to behave.

Learning processes immediately fol-lowing an animal's birth have crucial implications for the subsequent develop-ment of social behaviour. Imprinting, a form of rapid and intense learning, is confined to the first hours of an animal's life. It is particularly important to survi-val in birds with precocial young — young born in an advanced state of development and immediately able to move around by themselves and feed.

Through imprinting the young bird learns to make particular responses to only one type of animal or object. As soon as it hatches, the tiny bird must immediately recognize the object "mother" and follow it if it is to avoid predators and find food. In normal circumstances its mother is the first thing the young bird sees, so it is easy for this rapid learning to go according to plan.

In fact, the imprinting response is so powerful that the bird reacts to anything it sees and hears soon after hatching and in experiments will follow a human or even a box with flashing lights. Imprint-ing also helps young develop correct social responses to other members of the same species; sexual imprinting helps a developing animal direct later repro-ductive behaviour toward the right potential mates.

The evolutionary benefits of imprint-ing are clear: the bird is much more likely to survive if it stays close to mother than if it wanders off on its own. Imprinting also ensures that in normal circumstances parents care only for their own offspring.

Like almost all behaviour, imprinting is part predetermined, part the result of specific experience. Its inherited aspects include the capacity to imprint, the "sensitive" period over which such imprinting can occur, and the "critical" period at which the attachment response is best developed. In ducklings, the sensitive period during which imprinting can occur lasts up to about 30 hours after hatching, but the response is at its strongest between 12 and 17 hours – the critical period. The actual object to which the young animal becomes imprinted at this time is a matter of chance.

Play, another type of behaviour seen in young animals, is easy to recognize but difficult to define. It is behaviour with no immediate function. It does not directly enable the animal to find a mate, build a nest or capture food. But far from being

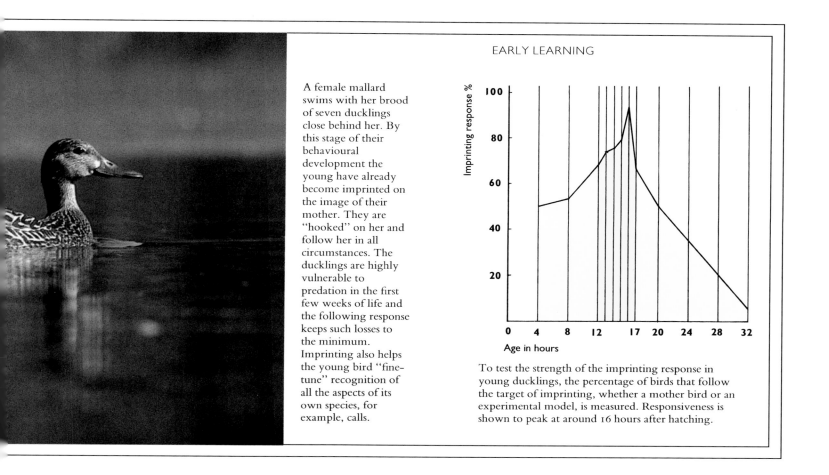

EARLY LEARNING

A female mallard swims with her brood of seven ducklings close behind her. By this stage of their behavioural development the young have already become imprinted on the image of their mother. They are "hooked" on her and follow her in all circumstances. The ducklings are highly vulnerable to predation in the first few weeks of life and the following response keeps such losses to the minimum. Imprinting also helps the young bird "fine-tune" recognition of all the aspects of its own species, for example, calls.

To test the strength of the imprinting response in young ducklings, the percentage of birds that follow the target of imprinting, whether a mother bird or an experimental model, is measured. Responsiveness is shown to peak at around 16 hours after hatching.

purposeless, play is actually a series of practice attempts at activities and skills that will be vital in adult life.

Mock attacks between young dogs or cats are rehearsals for later attacks on prey. A gambolling foal or lamb is developing muscle strength and coordination that will improve its chances when it needs to outrun an attacking predator. A young chimpanzee probing and poking every object in its environment is building up memory experiences that will help develop its foraging skills as an adult and thus ultimately increase its evolutionary fitness.

When a young lion cub pounces on its mother (*right*) or stalks a scuttling spider it is getting essential practice for hunting when it is adult. Such playful exuberance is in fact an adaptive strategy. Through play the young predator can experience a range of hunting scenarios without any risk of danger, and learn appropriate lessons from them.

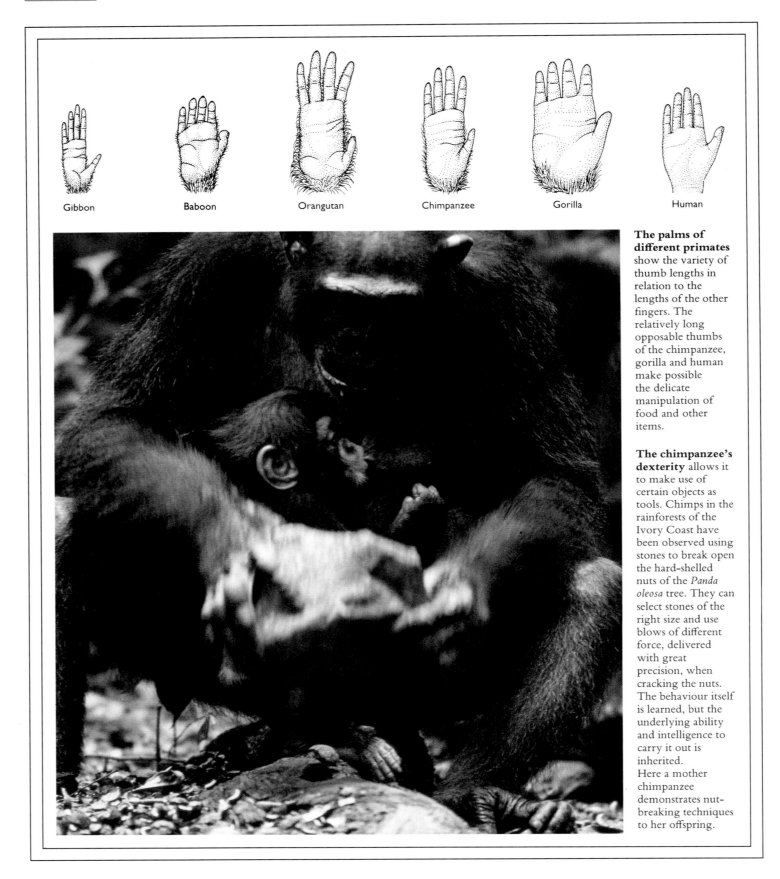

| Gibbon | Baboon | Orangutan | Chimpanzee | Gorilla | Human |

The palms of different primates show the variety of thumb lengths in relation to the lengths of the other fingers. The relatively long opposable thumbs of the chimpanzee, gorilla and human make possible the delicate manipulation of food and other items.

The chimpanzee's dexterity allows it to make use of certain objects as tools. Chimps in the rainforests of the Ivory Coast have been observed using stones to break open the hard-shelled nuts of the *Panda oleosa* tree. They can select stones of the right size and use blows of different force, delivered with great precision, when cracking the nuts. The behaviour itself is learned, but the underlying ability and intelligence to carry it out is inherited. Here a mother chimpanzee demonstrates nut-breaking techniques to her offspring.

USING TOOLS

The young chimpanzee's apparent playful curiosity sometimes leads it to experiment with potential tools such as sticks and stones and to try to copy its elders in using them. A tool is an inanimate object found and sometimes modified by the user and employed to carry out a specific task, such as getting food. An inherited trait for tool use in some individuals would give a distinct evolutionary advantage over non-tool users of the same species. Gradually, that inherited trait becomes more common until it is a feature of the species.

The Egyptian vulture is an expert tool user. The bird drops or throws a heavy stone at large eggs, such as ostrich eggs, to crack them open and obtain the protein-rich food. Sea otters open clams by banging them against heavy rocks. Without the "tool" the nourishing food inside the clam would be inaccessible to the otters. In both the vulture and the otter their survival chances are probably improved by their tool-using skills.

On the Galápagos Islands in the Pacific live 14 different species of finch that are thought to have descended from a common ancestor. First described by Charles Darwin, these finches played an important part in the development of his theories on evolution.

The different species of Galápagos finch now fit many of the roles filled by specialized and unrelated birds in other continents. One example is the woodpecker finch. In the absence of any woodpecker-like competitors, one population of the ancestral species evolved a daggerlike beak similar to that of the woodpecker. Like the woodpecker it uses this to dig into soft wood and cacti for larval insect prey. But the finch does not have a long woodpecker-like tongue to remove the exposed insects. Instead it holds a cactus spine in its beak to do the job of the tongue. Hence evolution has fitted the finch for its role by both structural and behavioural adaptations.

Chimpanzees find, use and modify a wider range of tools than any other non-human animal to gain access to food that would otherwise be difficult or even

Shaped rocks – a flake and a core – belonged to a prehistoric "tool kit" used by Neanderthal humans, who first evolved about 150,000 years ago. The evolutionary success of humans has depended in large part on our ability to use tools to a greater degree than other animals. This behavioural adaptation has gained us immense ecological advantages.

impossible to acquire. They employ the tools in a variety of ways. Many trim twigs and sticks and use them to extract termites from termite mounds. Others take a sharpened, chisel-like stick to open up a bees' nest, then puncture the nest lining with a thinner stick before extracting the honey with a flexible length of green vine. A wad of chewed leaves serves as a sponge to get water out of inaccessible holes in trees, while rocks and chunks of wood are used as hammers and anvils to crack thick-shelled nuts.

Swiss zoologists Christophe Boesch and Hedwige Boesch-Achermann have observed and filmed chimpanzee mothers teaching their young to use these tools. They believe that the young chimps do not fully master these skills until they are about six years old.

Humans have, of course, refined and expanded the making and use of tools to a unique extent, far beyond the capabilities of any other animal. But the abilities of chimpanzees reveal the potential for such behaviour among the apes. Our hominid ancestors seem to have developed in tropical African ecosystems already well stocked with efficient predators of mammalian prey. Tool use may have played a vital part in our success in competing with other carnivores and given us an evolutionary edge.

Coordinated family or clan use of stone hunting tools must have given early humans an advantage in fights over carrion with predators such as sabre-toothed cats and spotted hyenas. From the simplest shaped rocks through an expanded "tool-kit" of ever more refined and specialized objects, human tools for food capture and dismemberment marked the beginning of our development as a technological species. Human tool use can now shape or destroy our planet.

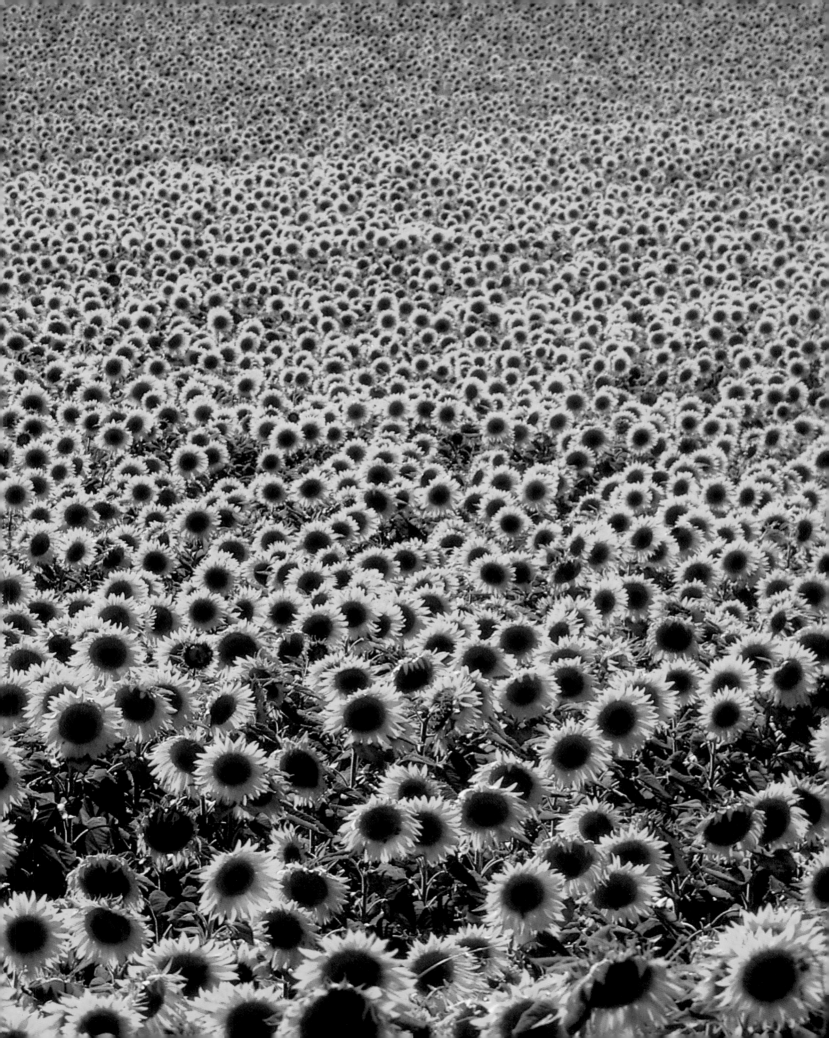

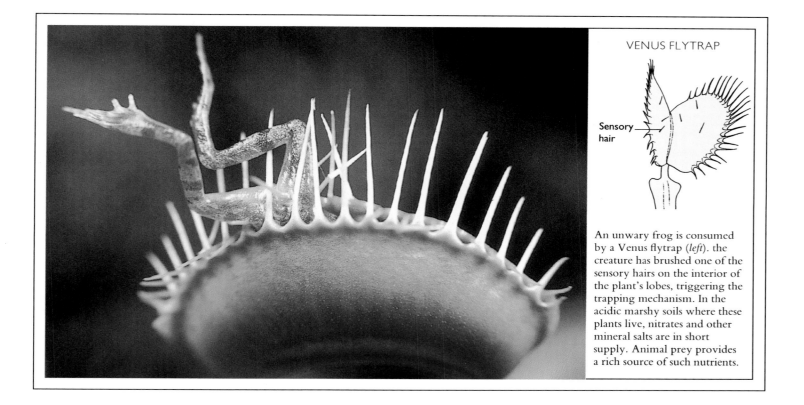

VENUS FLYTRAP

Sensory
hair

An unwary frog is consumed
by a Venus flytrap (*left*). the
creature has brushed one of the
sensory hairs on the interior of
the plant's lobes, triggering the
trapping mechanism. In the
acidic marshy soils where these
plants live, nitrates and other
mineral salts are in short
supply. Animal prey provides
a rich source of such nutrients.

ACTIVE PLANTS

Plants cannot "behave" in the complex ways of a tool-using chimpanzee, but they can and do respond to the world around them. Plants have two major guidance systems that steer their growth. Growing tips of roots or stems are sensitive to both the direction of gravitational pull and the direction of sunlight and help the plant orientate its growth appropriately – stems grow toward the sun, roots toward the pull of gravity. Many of these growth responses are driven by changes in the concentration of hormones in the plant. Although small, these changes accumulate over minutes or hours to produce significant behaviour. The way sunflowers all turn their heads to track the sun is an example of this sort of gradual action.

But some plants can also react to

Plants such as sunflowers keep their heads turned toward the sun. This "sun-following" behaviour helps keep the visual signalling of petals obvious to pollinating insects. The sun's warmth on the flower also speeds the growth of reproductive organs and the setting of seeds.

stimuli much more quickly, in ways akin to the rapid responses of animals. Carnivorous plants, for example, actually trap and consume living creatures. Most of these plants live in nitrogen-poor soil and may have evolved this behaviour mechanism to make up the nitrogen shortfall with nutrients contained in animal prey.

The Venus flytrap, *Dionaea muscipula*, has a typically ingenious mechanism. The outer portions of its leaves consist of two fleshy lobed sections with a central hinge. Each lobe is fringed with interlocking spikes and on its inner surface are three sensory hairs.

When the plant is in its trapping position, with the lobes apart, the hairs are easily brushed by any insect alighting on the lobe. These hairs are touch transducers – they can convert touch into electrical changes. If an insect touches a hair, the sensory cell at its base is squeezed. This produces a change in the electrical state of the sensory cell, which is rapidly transmitted to cells in the hinge zone. Normally kept pumped up with cell sap, the hinge cells instantly collapse when they receive the message that a hair

has been touched. Their collapse slams the trap shut so quickly that the intruding insect is captured.

To guard against one-off stimuli such as raindrops or wind-blown particles triggering the trap, there is a further piece of evolutionary fine-tuning to this behaviour mechanism. The trap is only triggered if two electrical messages from any of the six hairs are transmitted to the hinge within a period of 30 to 40 seconds.

The trap can also distinguish between trapped food, containing valuable minerals and amino acids, and nondigestible objects. If useful food is squeezed between the lobes, the trap stays shut and enzymes are secreted by glands on the lobe surface to digest it. If the plant senses that the trapped item is not food, the hinge cells reinflate and the trap opens.

Similar responses power the reproductive behaviour of flowers such as the hammer orchid, *Drakea*, of the dry grasslands of South Australia. If a wasp lands on its petals, the plant responds by throwing the insect at its pollen packets. Thus it ensures pollination.

SALMON MIGRATION

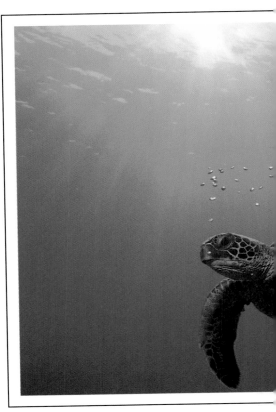

Like many salmon species, Atlantic salmon, *Salmo salar*, are born in fresh water and later migrate to rich marine feeding grounds. They then return to the rivers of their birth to spawn and the cycle starts again. The salmon seem to need the shallow, gravel-bottomed conditions of the rivers for the development of their eggs. The migration out to sea may be a behavioural adaptation to gain access to food resources.

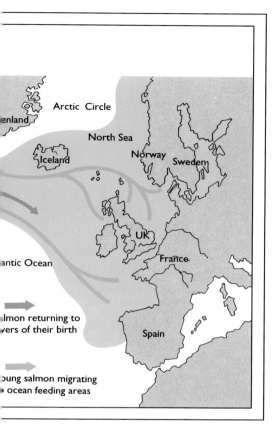

Arctic Circle

...enland

North Sea

Iceland   Norway   Sweden

UK

...antic Ocean   France

Spain

...lmon returning to
...vers of their birth

...oung salmon migrating
... ocean feeding areas

MIGRATION BEHAVIOUR

Among the most spectacular of all behaviour patterns are the extraordinary migratory journeys of some animals. Creatures such as many birds, fishes, turtles, whales, caribou and even butterflies periodically travel huge distances from one place to another, typically with a different climate and conditions. Often the migration is two way and animals may make many such journeys in a single lifetime.

These migrations – salmon unerringly returning from the ocean to their home spawning river, a green turtle travelling thousands of miles across the sea to find its natal beach – pose two evolutionary questions. First, why is such an energy-expensive activity such as migration an evolutionary advantageous trait? And second, what are the behaviour adaptations that allow such feats of navigation to come about?

An obvious motivation for migrators is to gain adequate food and a good climate throughout most of the year. Many migratory birds spend their summer and breed in the far north or south temperate zones, then migrate toward warmer equatorial zones to escape the cold and food shortages of winter. Humpback whales breed in the warm waters near the equator, but migrate back to cooler nutrient-rich waters near the poles to feed. Presumably the advantages of such food sources outweigh the disadvantages of the high energy cost of travelling to reach them.

The navigational abilities of migrating animals are legendary but still not really understood. Tiny olfactory clues carried in water may point the way for some aquatic creatures. The sun by day and the star pattern at night can provide a basic compass system for airborne migrators. Visually remembered components of a landscape help map the journey. Low-frequency sounds travelling over great distances may identify the

Green turtles, *Chelonia mydas,* travel huge distances to certain beaches where they lay their eggs. Some migrate more than 1,000 miles (1,500 km) from the coast of Brazil, where they feed, to a nesting site on Ascension Island.

direction of the nearest land for a bird flying over mid-ocean. Patterns of polarized light in the sky can orientate insects to the sun even when the sun is not visible, and there is increasing evidence that a sensitivity to the Earth's magnetic field may direct the journeys of some migratory animals.

Reseach on loggerhead and green marine turtles has revealed some clues about their navigation behaviour and shows how different sensory systems combine in their effects to produce complex migratory behaviour. Turtle eggs are laid in pits dug on beaches. The young hatch at night and move away from the darker, higher horizon provided by the dunes or woodland backing the beach toward the sea. Eventually they reach the waves. Turtles leaving Florida's east coast take an eastward route. This seems to be partly determined by a "wave compass" – the waves come principally from an easterly direction in the hatching period and the young turtles swim into the wave direction. Experiments have also revealed a magnetic component to the navigation. In the absence of other clues hatchlings swam east. If the orientation of the magnetic field was reversed, however, they swam in the opposite direction.

Once out into the Gulf Stream, the turtles swim into the huge clockwise current circulation of the North Atlantic where they remain until they reach sexual maturity and have to return to their home beach to lay eggs. The navigational methods used for this part of their complex migration are not known, but may include responses to dissolved odours in the water of the home coast.

Viewed from an evolutionary perspective, migrations are simply one of several ways animals adapt to the rigours of seasonal climates. Faced with predictable months of cold weather and poor food supplies, natural selection has pushed some species to the solution of staying put and hibernating, others to changing their food source. Migration overcomes the environmental problems by the dramatic response of, if necessary, travelling halfway round the world.

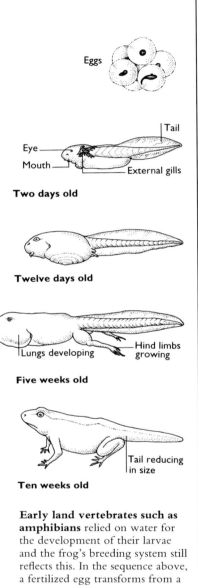

Eggs

Eye

Mouth

Tail

External gills

Two days old

Twelve days old

Lungs developing

Hind limbs growing

Five weeks old

Tail reducing in size

Ten weeks old

Early land vertebrates such as amphibians relied on water for the development of their larvae and the frog's breeding system still reflects this. In the sequence above, a fertilized egg transforms from a tadpole to a juvenile frog over a period of 10 weeks. At its earliest stages the tadpole is fishlike, with a propulsive tail for swimming and external gills for breathing under water. As it grows, the gills are lost and lungs develop to enable the frog to breathe air on land. As the tail reduces in size, limbs for movement on land appear.

ADAPTATIONS OF REPRODUCTION

All the organisms on our planet, from bacteria to insects, from sunflowers to mammals, share the same vital raison d'être – the need to reproduce themselves. Indeed, living things often appear to have the sole purpose of producing offspring and they exhibit the most ingenious strategies to achieve it.

During reproduction, individual creatures normally pass on to the next generation the genes that determine and control their traits and characteristics. Adaptive traits that improve reproductive success – measured as the number of new organisms in the new generation that survive, in turn, to reproduce themselves – will spread in a population. Traits that reduce this success are destined to disappear sooner or later. Ultimately, all adaptations become established if they improve an individual's chances of producing offspring that survive to reproductive age.

Adaptations to the mechanisms of the reproduction itself are geared toward achieving two distinct but interconnected ends. First, that greater numbers of offspring are produced and, second, that these offspring survive to reproduce themselves. No advantage is gained from simply producing large numbers of young. Only when such offspring live long enough to reproduce themselves will they contribute substantially to the persistence of a particular gene or species.

The best strategy for a species to achieve reproductive success will depend on habitat, competition and the physical limitations of its reproductive system. The strategy will be somewhere in a broad spectrum of possibilities. Ecologists have labelled the extremes of this spectrum "r" and "K". Reproductive strategies leaning toward the "r" extreme make many young as quickly as possible. This usually means producing huge numbers of tiny offspring and leaving them poorly provisioned and unprotected. Of these only a tiny fraction – but still an adequate absolute number – will survive to reproduce themselves.

Reproductive strategies leaning toward the "K" extreme aim to make small numbers of larger, well-nurtured and fiercely protected young, as in big cats and humans, for example. Each individual in such a brood, because it is vouchsafed a large investment of energy and care by its parents, has a correspondingly higher chance of surviving to a

The female arrow poison frog, *Dendrobates pumilio*, goes to great lengths to ensure the survival of her young. She carries her tadpoles on her own back up to pools of water trapped in bromeliads high in the jungle canopy.

reproductive age. The final outcome of both these "plans for reproduction" is identical — the production of enough surviving, reproducing offspring at least to replace the parental generation.

Closely related species inheriting dissimilar evolutionary histories and living in different environments may adopt contrasting reproductive strategies. Frogs typically produce thousands of small unprotected tadpoles from a mass of eggs, but only a tiny fraction survive to breed themselves. In the jungles of Costa Rica, however, female arrow-poison frogs produce a few large tadpoles which, by attaching themselves to her back, share the protection afforded by her warning coloration. These frogs have moved in the direction of the "K" extreme in the spectrum of possibilities.

THE MARSUPIAL STRATEGY

A far greater investment in the care and protection of individual offspring features in the reproductive strategy of all mammals. Female mammals (with the exception of egg-laying monotremes such as the platypus) carry their developing young within their bodies and provide them with nutrients. Since the end of the Cretaceous, some 65 million years ago, this internal and survival-enhancing life-support system has evolved along two independent tracks, as exhibited by both the marsupial and placental mammals.

The reproductive adaptations of marsupials centre, at least in the wallabies and kangaroos, on three staggered generations of offspring which live simultaneously off the same mother. Ecologists suggest that this effective strategy provides, on the one hand, a powerful insurance policy against harsh environmental changes and, on the other, the ability to expand the population rapidly when favourable environmental conditions return. This back-to-back pair of advantages partly explains the phenomenal success of marsupial mammals in the arid habitats of much of Australia.

THE PLACENTA

Female placental mammals, such as mice, monkeys and elephants, make a similarly

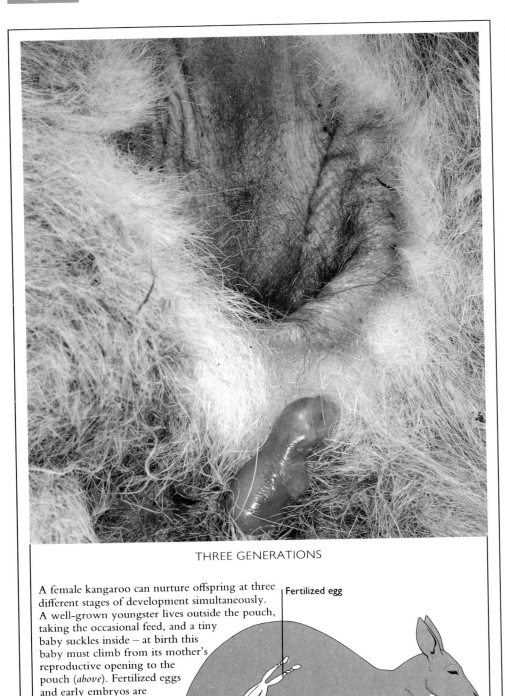

THREE GENERATIONS

A female kangaroo can nurture offspring at three different stages of development simultaneously. A well-grown youngster lives outside the pouch, taking the occasional feed, and a tiny baby suckles inside — at birth this baby must climb from its mother's reproductive opening to the pouch (*above*). Fertilized eggs and early embryos are retained in the oviduct as a third potential generation.

Fertilized egg

Baby in pouch

Young kangaroo

Almost immediately after birth a baby elephant must be up on its feet and able to move with the herd. It is born at an advanced stage of development having been nourished inside the mother's body for about 22 months through the placenta.

massive commitment but the pattern is radically different. Soon after fertilization, the egg, instead of developing freely in an oviduct, becomes embedded at an embryonic stage in the wall of its mother's uterus, or womb. At the point of contact a placenta, richly endowed with blood vessels, develops from maternal and embryonic tissues.

Although the blood of the mother and offspring never mix, the two sets of blood vessels come close enough to

A LIFE-SUPPORT SYSTEM

A key factor in the success of placental mammals, from mice to humans, is their ability to nurture and protect their offspring during long periods of development inside the mother. The baby lies in the uterus, cushioned by amniotic fluid. The umbilical cord connects it to the placenta – its life-support system. Via the placenta the baby receives nutrients and oxygen from its mother's blood and waste products such as carbon dioxide are removed.

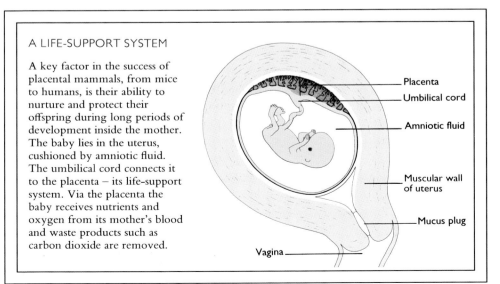

Placenta

Umbilical cord

Amniotic fluid

Muscular wall of uterus

Mucus plug

Vagina

ensure that nutrients and oxygen pass into the fetus, while its waste products are effectively removed by the mother. In addition a series of membranes, which enclose a layer of amniotic fluid, protectively surround the developing fetus. After they are born the offspring of placental mammals stay close to their mother and feed on her milk. During this last, often lengthy, phase the vulnerable youngsters can be physically protected by related adults until they become independent and are well on their way to reproductive adulthood.

SHELLED EGGS

The birth of live, partly independent young is a rarity in vertebrates other than mammals. Female fish, amphibians, reptiles and birds normally lay eggs. The females of most fish and amphibians lay eggs which are immediately fertilized externally by the sperm of one or more males. In birds and reptiles, however, eggs are fertilized inside the female's body after copulation and then coated with a hard drought-resistant shell.

A female bird lays a small number of tough, nutrient-rich eggs in one breeding season. Only a few species lay more than 20 or 30. Associated with this is the fact that only the left ovary and oviduct of the female's reproductive system is

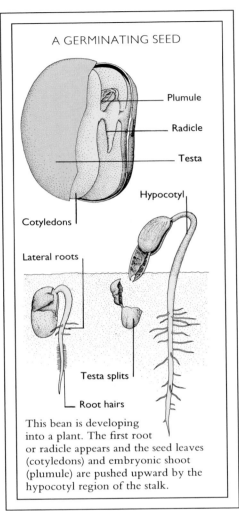

A GERMINATING SEED

Plumule

Radicle

Testa

Hypocotyl

Cotyledons

Lateral roots

Testa splits

Root hairs

This bean is developing into a plant. The first root or radicle appears and the seed leaves (cotyledons) and embryonic shoot (plumule) are pushed upward by the hypocotyl region of the stalk.

functional. Unfertilized eggs, or oocytes, are among the largest single cells known in the animal world. Each has a single gamete nucleus and is bulked out – to a size that may be more than an inch (2–3cm) in diameter in a large bird, such as the ostrich – by rich protein and fatty nutrients.

After fertilization in the oviduct the egg is surrounded by albumen – a protein-rich foodstore additional to that of the yolk – and enclosed in a hard mineralized shell. Minute pores in the shell allow oxygen to diffuse into the egg and carbon dioxide out. The egg is a remarkably self-contained and protected growth chamber that enables the fertilized egg to grow into an air-breathing chick at the time of hatching.

Shelled eggs were vital to the progress of evolution. The transition from a vertebrate way of life in aquatic habitats to one emancipated from water and able to diversify fully on the land is founded squarely on these seemingly humble reproductive products. Fish and amphibians had to lay their eggs in water. Reptiles coupled internal fertilization with drought-resistant eggs early on in their evolution, thereby enabling them to become the first thoroughgoing terrestrial vertebrates. They then passed on this pair of reproductive characteristics

THE KIWI'S EGG

The kiwi lays an egg that is 25 percent of its own weight – this X-ray shows just how much of its body is taken up by the egg. Kiwis are descended from moas, the biggest birds ever known, and large eggs may be an inherited characteristic.

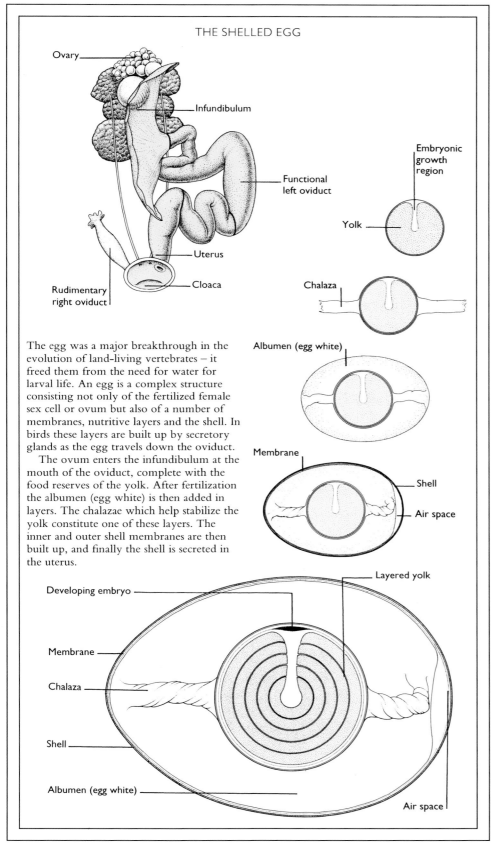

THE SHELLED EGG

Ovary

Infundibulum

Functional left oviduct

Uterus

Rudimentary right oviduct

Cloaca

Embryonic growth region

Yolk

Chalaza

Albumen (egg white)

Membrane

Shell

Air space

Developing embryo

Layered yolk

Membrane

Chalaza

Shell

Albumen (egg white)

Air space

The egg was a major breakthrough in the evolution of land-living vertebrates – it freed them from the need for water for larval life. An egg is a complex structure consisting not only of the fertilized female sex cell or ovum but also of a number of membranes, nutritive layers and the shell. In birds these layers are built up by secretory glands as the egg travels down the oviduct.

The ovum enters the infundibulum at the mouth of the oviduct, complete with the food reserves of the yolk. After fertilization the albumen (egg white) is then added in layers. The chalazae which help stabilize the yolk constitute one of these layers. The inner and outer shell membranes are then built up, and finally the shell is secreted in the uterus.

to their bird descendants. Without shelled eggs the line of land-living animals that produced reptiles, mammal-like reptiles, mammals, primates and ourselves could not have been sustained.

SEEDS AND FRUIT

The coated seeds and fruits of flowering plants are in many ways the "green" equivalents of animal eggs – the seed inside a fruit is a protective life-support system for a pollinated ovule. The outer coat, or testa, of the seed is physically tough and able to resist desiccation. Within the testa the developing embryo is nourished by carbohydrate, protein and fat stores. In plants known as mono-cotyledons, which include grasses and arum lilies, the nutrients are stored in a single cotyledon, or seed leaf; in many of the other common plant families, which form a group called dicotyledons, the stores are located, as their name suggests, in a pair of cotyledons.

Before a seed has produced roots to obtain water and mineral salts, and consequently spread leaves to trap light and obtain carbon dioxide, useful photo-synthesis is impossible. Cotyledon food stores tide the young plant over until photosynthesis enables it to become self-sustaining.

SEED DISPERSAL

Many plants and a few animals, such as barnacles, that live out their lives fixed to one spot (or are able to move only a short distance) often rely on their offspring to act as dispersal agents. If successful, such dispersal can extend the range of a species, while at the same time prevent-ing overcrowding and the exhaustion of limited resources close to the parents. Moreover, dispersal facilitates the mix-ing of an extended gene pool, thereby increasing the chances that sexual repro-duction might bring together two other-wise separate advantageous mutations.

The small products of reproduction are ideally suited to move or be moved over short, medium or long distances. Fungi, for instance, make prodigious numbers of tiny spores, each only a few thousandths of a millimetre across. A puffball fungus may contain billions of

The impact of a raindrop stimulates the puffball fungus (*left*) to release clouds of spores into the air. The few that settle in the right habitat will germinate, producing a new fungal mass (mycelium) and eventually a new puffball.

Wind dispersal ensures that the seeds of the dandelion (*above*) are carried away from the parent plant. Each mature seed on the flowerhead has a stalked "parachute" of fine hairs which is easily caught by the wind.

spores. A breath of wind or the impact of a water droplet causes a mature puffball to release a cloud of spores that become dispersed over many miles.

The bewildering diversity of seed dispersal mechanisms found in flowering plants are either mechanical devices or adaptations that exploit wind, water or animals. The fruit of gorse and of some species of *Impatiens*, for example, have catapult devices that in dry conditions "shoot" seeds many feet from the parent. Maple and sycamore seeds have wings that rotate like helicopter blades in the wind, while a dandelion seed is endowed with a hairy parachute that carries it through the air. Coconuts, with built-in protection from excessive salt, are adapted to float for long periods in the sea, enabling them to "island hop" throughout the tropics. The seeds of geums and burdocks have "hooks" that stick like Velcro to the fur of a mammal and are dispersed as the creature moves.

INTERNAL PARASITES

The fraction of their bodies and nutrients that most animals can divert toward reproduction is a delicate and ever-changing "trade off" with other demands. To survive at all an animal must first feed and defend itself from predators. Only then can it divert resources into reproductive activities.

However, one group of animals – the internal parasitic animals known as endoparasites – concentrate overwhelmingly on reproduction. Creatures such as flukes, tapeworms and parasitic nematodes flourish in a superabundance of food provided by the body or gut contents of their host. Living in the gut, blood system or other organs of a host, parasites do not have to devote resources to finding food or defending themselves from predators. Indeed, apart from attack by the immune defences of their

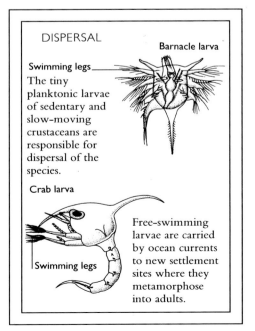

DISPERSAL

Barnacle larva

Swimming legs

The tiny planktonic larvae of sedentary and slow-moving crustaceans are responsible for dispersal of the species.

Crab larva

Free-swimming larvae are carried by ocean currents to new settlement sites where they metamorphose into adults.

Swimming legs

host, endoparasites experience few negative features from their immediate environment. Consequently, most become hyperactive reproducers, turning host-provided nutrients into offspring.

A female worm of the human intestinal nematode *Ascaris*, for instance, produces 240,000 eggs a day. Since each worm lives for about a year, its lifetime's reproductive effort produces nearly 100 million eggs.

The human blood fluke *Schistosoma*, which infects over 200 million people worldwide, follows a more complex reproductive strategy than *Ascaris*. In its convoluted amalgam of sexual and asexual reproduction, *Schistosoma* has capitalized on a number of remarkable adaptive opportunities. Adult males and females live together in copulating pairs inside human blood vessels (see diagram). Worms of some *Schistosoma* species congregate around the intestine, those of others close to the bladder. About half of the constant stream of spiny eggs that the female releases end up in tissues all over the host's body where they cause the damaging symptoms of the disease schistosomiasis, or bilharzia.

The other half of the eggs damage the walls of the host's gut or bladder and enter its feces or urine. In Africa, Asia or South America where the disease is prevalent, these eggs often end up in freshwater streams and lakes. Here, ciliated larvae hatch from the eggs and bore into snails. Eventually, after several stages of asexual multiplication, microscopic larvae called cercariae leave the snail and enter the water. They penetrate the unprotected skin of anyone who swims, works or washes in the infected water and become adult worms.

The mixture of asexual and sexual reproduction in the same life cycle has undoubtedly contributed to the success of the *Schistosoma* parasites. The asexual cloning in the snails is the most economical way of producing huge numbers of genetically identical offspring quickly (pp. 72–73). The sexual phase in human blood promotes, through gene shuffling, genetic diversity and is probably the reason why six or seven different species of human schistosomes exist today.

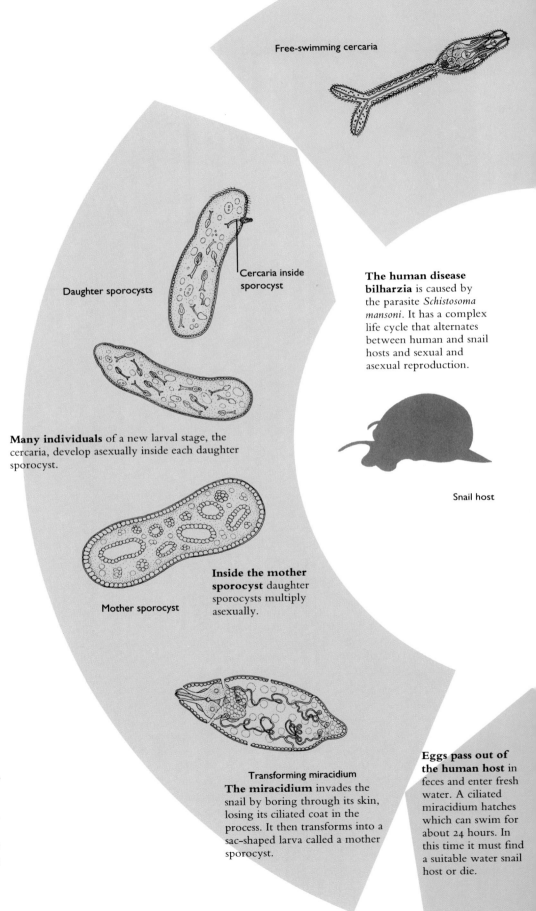

Free-swimming cercaria

Daughter sporocysts

Cercaria inside sporocyst

Many individuals of a new larval stage, the cercaria, develop asexually inside each daughter sporocyst.

The human disease bilharzia is caused by the parasite *Schistosoma mansoni*. It has a complex life cycle that alternates between human and snail hosts and sexual and asexual reproduction.

Snail host

Inside the mother sporocyst daughter sporocysts multiply asexually.

Mother sporocyst

Transforming miracidium
The miracidium invades the snail by boring through its skin, losing its ciliated coat in the process. It then transforms into a sac-shaped larva called a mother sporocyst.

Eggs pass out of the human host in feces and enter fresh water. A ciliated miracidium hatches which can swim for about 24 hours. In this time it must find a suitable water snail host or die.

Thousands of cercariae emerge from the snail into the river. They then penetrate the skin of people standing or swimming in these waters. The forked swimming tail of the larva is shed during this invasion. The remaining head region transforms into a young adult.

Human host

A pair of adult worms is joined in permanent copulation inside the blood vessels of a human intestine.

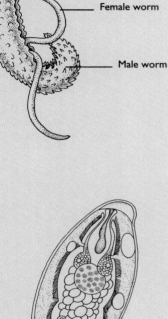

Female worm

Male worm

Spined egg

This spined egg is one of the thousands produced by the copulating worms. It contains a ciliated larva called a miracidium. The spine helps the egg escape from blood vessels into the human gut.

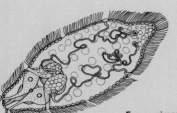

Free-swimming miracidium

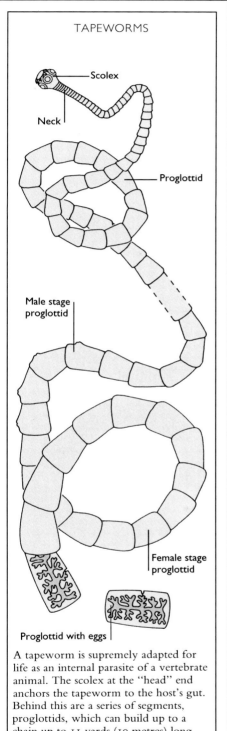

TAPEWORMS

Scolex

Neck

Proglottid

Male stage proglottid

Female stage proglottid

Proglottid with eggs

A tapeworm is supremely adapted for life as an internal parasite of a vertebrate animal. The scolex at the "head" end anchors the tapeworm to the host's gut. Behind this are a series of segments, proglottids, which can build up to a chain up to 11 yards (10 metres) long. Each segment develops first a set of male reproductive organs and then a female set. Male stage segments fertilize segments at the female stage of development which then change into sacs each containing thousands of eggs.

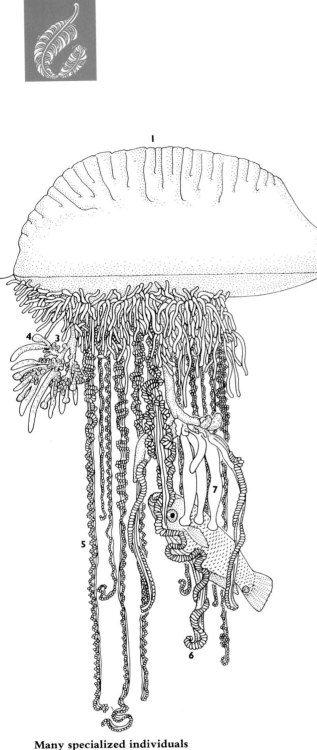

Many specialized individuals
make up the Portuguese man of
war, a tight-knit colonial
organism. They include the sail
(1), part of the float individual (2)
which keeps the colony on the
water surface, and male (3) and
female (4) reproductive forms.
Other individuals are specialized
for capturing prey and the
digestion of food. Here, elongate
(5) and contracted (6) catching
tentacles hang down beneath the
float while feeding individuals (7)
consume a captured fish.

ADAPTATIONS OF SOCIAL ORGANIZATION

One of the most fascinating aspects of life on Earth is the way in which closely related individuals within a species live together in closely knit societies. And, just like the individuals of which they are composed, such societies are themselves the "raw material" for evolution – they may adapt and change over time in response to alterations in their circumstances. The results of such evolution have created the complex societies of animals such as ants, bees and wasps.

It is not always easy, however, to determine the circumstances in which an individual must be regarded as a member of such a society. The boundary between organisms that are discrete and separate, such as individual humans, fish and maize plants, is clear and obvious. However, when closely related organisms are in some way linked together, the boundary between individuals becomes blurred.

A coral on a reef, for instance, is composed of many polyps which, at night, spread out their tentacles to catch food. As they do so, they resemble a multitude of individual sea anemones emerging from the rock surface. In fact, the polyps are linked by a communal gut beneath the rocky portion of the coral, so that food captured by the tentacles of one polyp is shared by hundreds of others.

A solitary sea anemone can legitimately be called an individual organism because it is not physically connected to other members of its species. The status of the polyps of a coral is more difficult to specify. Is the coral one super-organism with hundreds of heads or a series of individuals that have been produced by asexual reproduction but have failed to separate completely? In a sense it is both, as is the related Portuguese man of war, in which different polyps are specialized for a range of particular functions.

The polyps of a coral are all genetic clones of each other so it is entirely appropriate for the polyps to share nutrients in a seemingly selfless, non-competitive way. Competition has been abandoned for the sake of the colony as a whole because any adaptation that fosters the survival and further reproduction of the coral will inevitably result in new genetic copies of each and every polyp.

The Portuguese man of war, *Physalia* sp., is a committee organism – a colony of hundreds of different individuals. The largest of these is the float, or pneumatophore, which lies on the water surface and keeps the colony afloat. Other individuals, specialized for tasks such as feeding or reproduction, live below the float. All are physically joined and genetically identical.

GENETIC RELATEDNESS

The colonial coral is an extreme and clear-cut example of the relationship between genetic similarity and adaptations of behaviour that involve cooperation in groups of organisms belonging to the same species. As a rule, the more genes shared by a cluster of organisms, the more related they are, the greater the kinship between them and the greater the prospect of apparent altruism and cooperation. As far as evolution is concerned these social behavioural traits will be adaptive if they help more genes get into the next generation than would be the case if no cooperative behaviour were to take place.

Many major groups of animals, such as insects, birds and even mammals, include species for which tightly knit colonies with considerable levels of social cohesion, joint cooperative activity and sophisticated within-group communication act as evolutionary "units" through which change and adaptation are expressed.

Whether a polyp consumes the food it catches or shares it with its genetically identical neighbours makes little difference to the reproductive fitness of the whole coral. But in colonies where individuals are not genetically identical, such as breeding colonies of birds, help and harm between animals can have more complicated consequences. Herein lies a central yet subtle consideration. How does an individual's particular way of behaving benefit or harm the reproductive fitness of either itself or the colony to which it belongs?

The best reproductive strategy for an individual animal in any circumstances should involve, it might be supposed, unbridled "selfishness". If an animal's overriding evolutionary purpose is to endow as many offspring as possible with a genetic constitution similar to its own,

Colonies of white-fronted bee-eaters (*right*) make their nest burrows on a cliff face. But only certain dominant male and female pairs in a colony breed. Each of these breeding pairs is aided in nest building, defence and feeding nestlings by a small group of related birds that, though fertile, do not breed in that season.

what could be more beneficial than concentrating every effort on personal reproduction? The answer is that, in certain conditions, an animal might be able to pass on more genes identical to its own by assisting another individual to breed rather than by breeding itself.

In this context, it is important to remember that in orthodox sexual reproduction – between diploid males and females (pp. 68–69) – a parent passes on only 50 percent of its genes to its offspring. Overall reproductive fitness must be a combination of this level of genetic relatedness and the likelihood of successfully producing offspring that themselves survive and breed.

TO BREED OR NOT TO BREED?

Imagine a hypothetical example in which female animal A is the sexual offspring of experienced parents B (the father) and C (the mother). The reproductive and parental experience of B and C, at the time their "fitness budgets" are analysed, indicate that they will produce offspring with a 60 percent chance of survival. The inexperience of A means that, if she breeds in her first adult year, her offspring have only a 20 percent chance of survival.

Which is A's best fitness strategy in these conditions? Should she breed herself or spend her energies in this breeding season helping her parents produce another brood? Self-breeding means that 50 percent of A's genes will pass into the next generation, but with only a 20 percent chance of survival. Thus, the total gene flow will be proportional to 50 percent of 20 percent, that is 10 percent.

If A forgoes reproduction herself and instead helps her parents to boost their infant survival rate from 60 percent to 100 percent, a different reproductive outcome ensues. On average A will share a 50 percent genetic identity with any

Teamwork helps meerkats survive in the Kalahari desert in Africa (*above left*). The meerkat, *Suricata suricatta*, is a type of mongoose and lives in highly social colonies of 20 to 30 animals. All adults within the group mate but they cooperate in the feeding and care of young and in the defence of the colony.

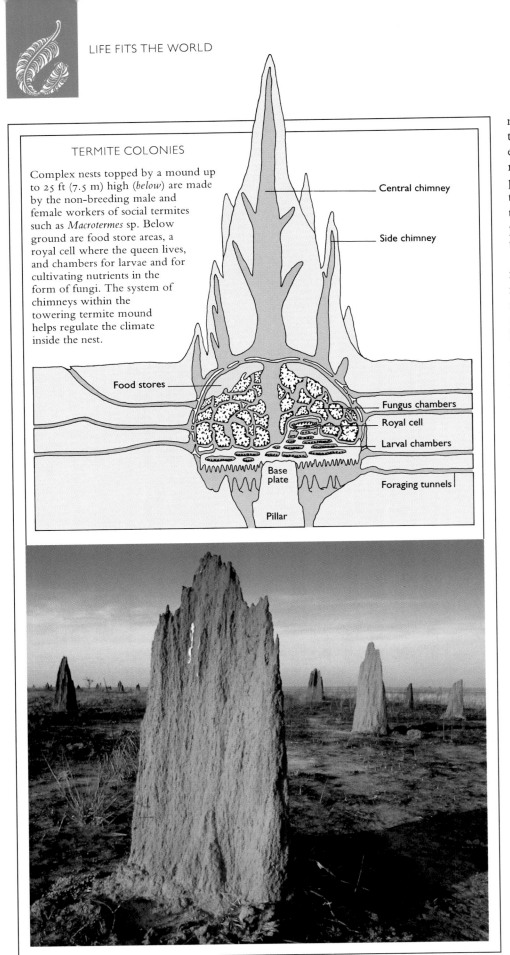

TERMITE COLONIES

Complex nests topped by a mound up to 25 ft (7.5 m) high (*below*) are made by the non-breeding male and female workers of social termites such as *Macrotermes* sp. Below ground are food store areas, a royal cell where the queen lives, and chambers for larvae and for cultivating nutrients in the form of fungi. The system of chimneys within the towering termite mound helps regulate the climate inside the nest.

Central chimney

Side chimney

Food stores

Fungus chambers

Royal cell

Larval chambers

Base plate

Foraging tunnels

Pillar

new offspring produced by B and C – they will be full siblings of hers. So, effectively, each new offspring will receive 50 percent of her genes and 100 percent of the brood will survive. Thus, the total gene flow will be proportional to 50 percent of 100 percent, that is, 50 percent, so more of A's genes pass into the next generation by this strategy.

This hypothetical but perfectly plausible example shows that female A should not breed in her first adult year. Instead, she will contribute more to her overall (sometimes called inclusive) fitness by helping her parents to reproduce.

WHITE-FRONTED BEE-EATERS

Examples of individual animals helping others to breed can be found in a variety of social animals, including bee-eaters and meerkats. The brilliantly coloured white-fronted bee-eaters are insectivorous birds with a rapid acrobatic flight. They are gregarious inhabitants of dry water courses, eroded gullies and scrub-covered stony hillsides in dry savanna areas of east and southern Africa.

Breeding colonies of 10 to 20 active nests, comprising in total over a hundred individuals, occupy burrows in a sandy cliff. During the breeding season, a small clan of helper birds supports the primary breeding pair at each nest by defending the nest burrow against other birds and snakes, and by providing food for the hatched young.

The helper birds behave in this apparently altruistic way because, genetically, they are closely related to the breeding pair, either as siblings or as offspring from a previous season. The brevity of the breeding season that follows the rains makes it difficult to rear young successfully. Problems of food supply and the attentions of predators mean that even a clan of five or six cooperative adults might only rear one youngster to reproductive age during a breeding season.

Given the high odds against successful breeding, it is a more fruitful evolutionary strategy for the helpers to assist experienced, closely related breeders than to attempt to breed themselves. Thus, real birds in the African savanna imitate the behaviour of the animals A, B and C discussed in the hypothetical example when selective imperatives push them in that direction.

SOCIAL COMMUNITIES

Most of the insect species that form fully social communities belong to Hymenoptera, a taxonomic grouping which includes the ants, bees and wasps. These eusocial communities – communities that show the highest level of adaptations for social life – may each number over a million individuals. In them the trend toward assisting closely related animals to breed reaches its apogee. At the heart of a eusocial community there is a solitary breeding female surrounded by workers that, typically, never breed at all. In the ants and their relatives these workers are all female.

A key feature of the hymenopteran eusocial insects, and one that seems to pre-adapt them for their extremely social life, is the unusual way sex is determined. Most forms are what is called haplodiploid – females have the full diploid set of chromosomes, while males, which grow from unfertilized eggs, are haploid with only half a set. Every cell in a hymenopteran male's body more or less resembles the chromosomal state of a normal animal's sperm.

In the more orthodox system of sex determination, as found in humans, males have two different sex chromosomes (X and Y) while females have two similar ones (X and X). Since females can only produce eggs with an X chromosome, the sex of any offspring is entirely decided by the sex chromosome in the fertilizing sperm from the male. An X sperm specifies female offspring, a Y sperm will produce males. Since males make equal numbers of X and Y sperm, any large population of offspring will be half male and half female.

Most individuals in a hymenopteran colony are female workers. These females are sterile because their reproductive development has been curtailed by chemical signals, or pheromones, emitted by a large, dominant, breeding queen.

The female workers are the direct progeny of the queen, who stores in her body sperm from a single male. They are

SOCIAL BEES

A honeybee colony is one large family. Three types of bee live in the colony: the queen (the only breeding female) which produces all the eggs, a few male drones destined to mate in the next generation with virgin queens, and thousands of non-breeding female workers. These workers are all sibling daughters of the one queen and 75 percent identical to each other genetically. They make up the bulk of the colony and perform all the building, food-gathering and other tasks.

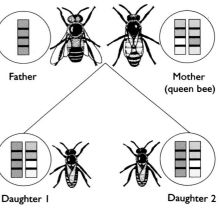

Father   Mother (queen bee)

Daughter 1   Daughter 2

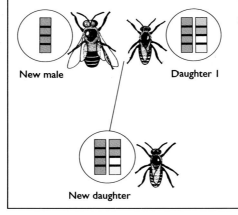

New male   Daughter 1

New daughter

Queens and female worker bees have two sets of chromosomes – they are diploid. But male drones are haploid – they have only one set. All sibling workers are 75 percent genetically identical – they share all the father's genes and half of the mother's. If a worker did mate, her female offspring would be only 50 percent genetically identical with her. It is therefore a more successful strategy for workers to maintain the colony and ensure the production of more siblings than to reproduce themselves.

Swollen with liquid nectar, these honeypot ants are like living storage jars. In several species of ant a special caste of non-breeding workers, known as repletes, has evolved. These ants are fed with so much nectar and honeydew by other workers that their abdomens become enormously distended and they can no longer move around. They spend their lives clinging to the roof of the underground nest.

In times of drought or food shortage the colony feeds on these reserves. The ants simply stroke the repletes, thus stimulating them to regurgitate the food.

WORKING TOGETHER

Small worker

Large worker

Leaf-cutter ants, *Atta* sp., feed on fungus which they grow on patches of chewed up leaves, mixed with fecal material, in their huge underground nests. Two different sizes of worker cooperate in the gathering of leaves. Large workers cut the leaves from plants and carry them to the nest while small workers perch on the leaves fending off the parasitic flies that lay their eggs on ants.

all her sterile, genetically identical daughters which make the colony operate successfully. Bee workers, for example, build the hexagonal cells of the comb out of wax secreted by their own bodies. Wasp workers manufacture papier-mâché pulp from plant material and then fashion it into a lightweight nest. Ant workers construct intricate tunnels and nests from plant debris.

Among their other tasks female workers collect all the food the colony needs, often from a great distance. Moreover, they can communicate the whereabouts of newly discovered food sources to other workers – flightless ants lay chemical (pheromone) trails on the ground and honeybees perform complex "dances" on their return to the hive. The precise pattern and orientation of the honeybee's dance tells other workers the nature, size, distance and direction of a new food source. Workers also defend the colony against attack, act as midwives to the queen, tend her eggs and provide her hatched young with food.

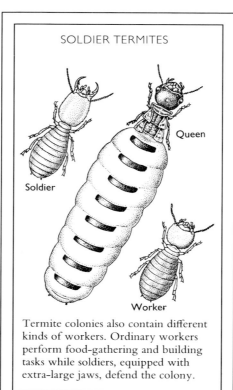

SOLDIER TERMITES

Soldier

Queen

Worker

Termite colonies also contain different kinds of workers. Ordinary workers perform food-gathering and building tasks while soldiers, equipped with extra-large jaws, defend the colony.

THE PERCENTAGE GAME

If every organism's most pressing imperative is to maximize its reproductive fitness why do all these hymenopteran females forgo offspring production? And why is such a strategy successful as a social adaptation? Again, the answers spring from the issue of relatedness. Hymenopteran sexual reproduction involves a fascinating distortion of normal kinship relations in which the "selfless" labour of the workers usually makes them "fitter" in evolutionary terms than they would be if they bred independently.

In a haplodiploid species, sister workers in a bee colony are on average 75 percent genetically identical (see p. 153). This is because half of their genes (those inherited from their haploid father) and half of the remaining half (those inherited from their diploid queen) are identical.

Every time female workers help the queen to produce more offspring they help to create new individuals that

A COLONIAL MAMMAL

Naked mole rats, *Heterocephalus glaber*, live in colonies in extensive underground tunnel systems beneath the sun-baked soil in parts of eastern Africa. These bizarre-looking creatures are only up to 4 in (10 cm) long, have baggy pink skin, almost no hair and two pairs of sharp, protruding teeth. They excavate a vast network of tunnels to reach the underground plant tubers on which they feed. The network can be huge. Some have been discovered that are almost 2 miles (3 km) long.

These tunnels are built by the workers in the colony. Some dig while others transport the earth to an opening at the surface. Just like insect workers, they also defend the colony and tend the offspring of the queen. In most instances these offspring are the full siblings of the colony's workers and therefore share their genes.

An average litter of pups is about 14 but it can be more than 20. The breeding queen has a specific adaptation which enables her to bear large litters. Once she starts breeding her body lengthens – the vertebrae spread out. This allows her to carry her large litters without becoming too fat to get through the tunnels.

In this colony (*far right*) workers are tunnelling (1) and clearing rocks from passages (5) while another mole rat gnaws at a tuber (2). Two other workers are trying to block off a tunnel which a predatory snake has entered (3). The queen lies in her chamber surrounded by offspring (4). There is a separate latrine area to which a worker escorts young (6). Life is not always cooperative, however, and aggressive battles do break out between members of the mole rat colony (7).

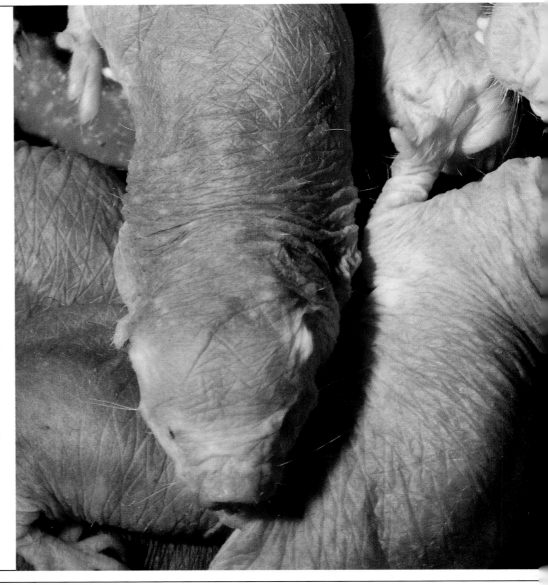

share 75 percent of their own genes. This is 25 percent better than any parent can manage in normal sexual reproduction. More to the point, however, is the fact that if a worker "selfishly" mated with another male her female offspring would, at best, be only 50 percent identical with her.

This 25 percent benefit seems to have had the effect, over evolutionary time, of turning many hymenopteran species into fully social animals. Mutations that produced variant behaviour patterns or prevented responsiveness to the queen's pheromone constraints on sexual development would be selected against because they would result in reduced reproductive fitness compared with the fully social mode.

NAKED MOLE RATS

At one of his lectures during the 1970s, the ecological theoretician Richard Alexander predicted the sort of characteristics that a eusocial mammal – if one were to be discovered – would have to possess. The mammal would be a small, rapidly breeding rodent which, because of its vulnerability to the outside world, lived permanently underground. A member of his audience declared that naked mole rats suited this description and subsequent studies have shown that these unique rodents are, of all mammals, the closest social equivalent to the bees and other eusocial insects.

The rats live in isolated colonies of a hundred or so individuals. Breeding is the preserve of a single female queen and one, or sometimes two, chosen males. The remainder of the colony's males and females act, in genetic terms, as the

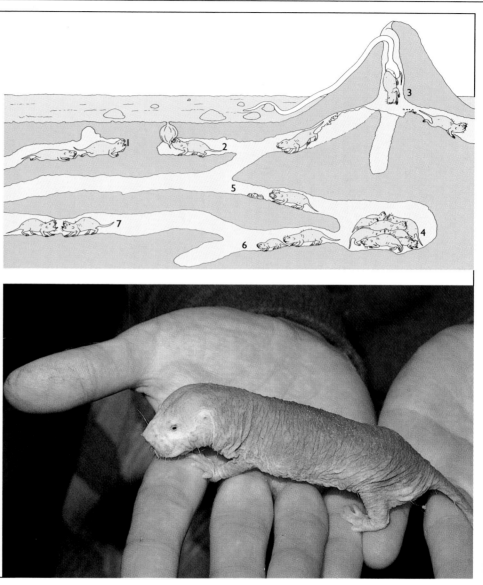

equivalent of bee or ant workers. They build the tunnel systems and defend the colony and young.

Naked mole rats are genetically diploid and sex is determined normally. So what selective advantage has moved them, reproductively speaking, in the direction of hymenopterans? Many scientists have come to believe that the answer lies in a combination of inbreeding and ecological features.

DNA "fingerprinting" reveals that the genes of individual mole rats in a colony are remarkably similar. In fact, the scale of inbreeding in a colony mirrors what would be expected in a human community if brothers and sisters were to inbreed for 60 generations. This genetic homogeneity means that, as with the hymenopterans, advantages will be gained from any adaptive behaviour by non-reproductive workers that helps the queen to produce more young. New offspring will carry copies of almost all of the genes in pre-existing workers.

The high levels of inbreeding arose, perhaps, through ecological pressures. Not only are colonies of naked mole rats isolated but their tuberous food grows in far-flung patches in an impossibly hostile environment. Since the rats cannot survive on the surface – they cannot even maintain their own body temperature – interchanges between colonies are rare. Once such high levels of inbreeding became established as the norm, the increasing genetic similarity of individuals in a colony would probably predispose them to eusocial reproductive behaviour. Because members of the colony share a high proportion of their genes, they help themselves by helping the colony.

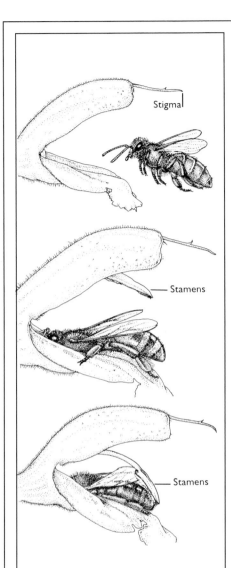

Stigma

Stamens

Stamens

A delicate trade-off exists between plant species and their pollinators. The plant must provide something to attract the pollinator and make its visit worthwhile. This reward is usually nectar – a nutritious sugary secretion – or pollen itself, both of which provide food for pollinating insects and birds.

This sage flower has particular adaptations that ensure its pollination. As a bee enters the flower to collect nectar its movements trigger the hinged stamens which swing down, brushing the bee with pollen. When the bee visits the next flower it leaves some of this pollen on the female stigma, so ensuring cross-pollination of the flowers.

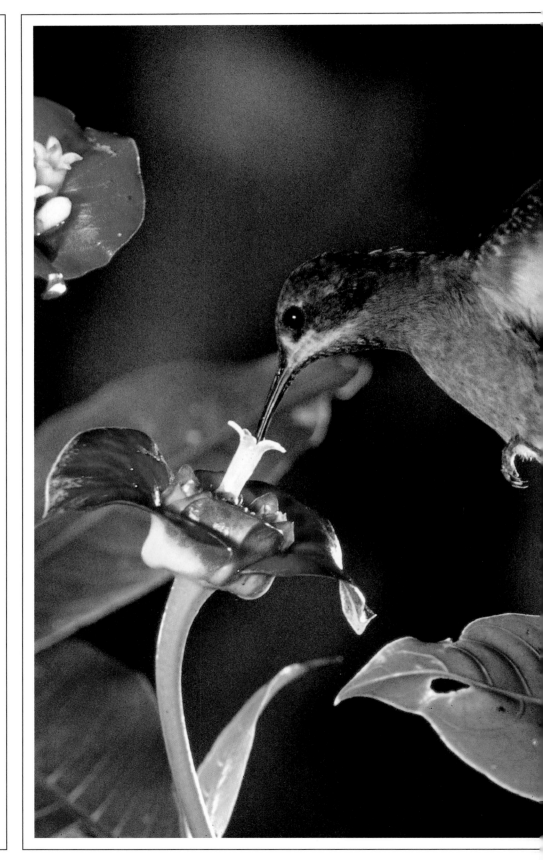

ADAPTATIONS OF COEVOLUTION

Just as "no man is an island", so all living things are affected in some way by the other sorts of organism with which they share our planet. In evolution, the effects of an association between two species can become permanently imprinted into the genetic constitution of one or both of them. The evolution of herbivores, manifested for example in the adaptive design of their teeth, is linked with the plants they eat. Similarly, plants have evolved such features as spiny leaves or the manufacture of toxic chemicals which defend them against being eaten by particular sorts of herbivore. In the world of predator-prey relationships, parallel stories abound.

All habitats are composed, at least in part, of living things, and all contain a community of species, each living and evolving side by side. The individuals of the different species interact with one another – by competing for resources, for instance, or else by simply eating each other according to their place in the food chain. It is in this context that living things adapt and evolve together as each improves its fitness for survival in an environment which, although changing, predictably contains other species.

When two species live closely together in these conditions, a change in one may stimulate an alteration in the other. Over long periods of time, therefore, a reciprocating pattern of adaptations unfolds. The two species no longer evolve separately but coevolve as an interlocked pair.

The most intriguing examples of coevolution are those in which the species spend large parts of their lives in close physical association, through parasitism or symbiosis (sometimes called mutualism).

In parasitism, one species – the parasite – lives on or in the body of another – the host. The parasite uses the host as its habitat and source of food but, in doing so, damages it. A fitness gain for the parasite means a fitness loss for its host. Coevolution in such an association often becomes a neverending "arms race". The host makes a defensive adaptation – perhaps in its immune system – to ward off the parasite, which then counters the defence. Again, the host responds defensively to the parasite and so the arms race continues through generation after generation.

A brown violet-ear hummingbird, *Colibri delphinae*, plunges its long beak into a flower to feed on nectar. Beak and flower structures have evolved in parallel to produce an efficient feeding system for the bird and an equally efficient pollinating mechanism for the plant.

By contrast, symbiosis improves the survival chances and the reproductive output of both partner species. In evolutionary terms, the partners are fitter living together than if they lived apart. All flowering plants, for instance, could employ wind or water to achieve pollination – the fertilization of one flower by pollen from another of the same species – and indeed many do. But some species have devised means of using animals to accomplish this crucial reproductive objective more effectively. The major groups of animals engaged in this type of symbiosis are insects and birds, which take pollen from one plant and transport it to another, while bats often pollinate plants with flowers that open at night.

Ever since flowering plants evolved during the Cretaceous they have offered pollinators the worthwhile rewards of nutrient-rich nectar, or pollen itself, in exchange for regular visits to their blooms. Moreover, plants such as the meadow sage (p. 158) have developed extraordinary mechanical devices to ensure that an alighting insect triggers precise movements of its flowers' reproductive organs.

ONE-TO-ONE DEPENDENCE

Partners in a pollination symbiosis may be utterly dependent on one another. For more than 65 million years fig trees have been locked into an intricate coevolutionary pact with the fig wasps (agaonids). Each species of fig has an exclusive partnership with a particular species of minute wasp. The fig seeds are pollinated only through the agency of the wasps. What is more surprising is that the wasps can only reproduce inside the fruits of their fig associates (see diagrams). Almost every aspect of the reproductive process of figs and fig wasps has evolved in response to selection pressures stemming from the reciprocal symbiotic partner. The fates of these two groups of organisms have become completely inseparable from one another.

A key feature of this most unusual pollination story is the fact that the female wasp actively collects pollen from male fig flowers and distributes it on female ones. Equally remarkable is the

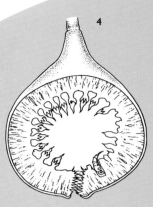

3 Wingless male wasps emerge from their pupae and inseminate females which are still in their pupal cases. By this stage the fig has grown an inner carpet of simple male flowers packed with pollen.

4 After copulating some male wasps die immediately. Others bore holes in the wall of the fig and then die.

Once pollinated, the long-styled flowers develop into fig seeds. The ovules of short-styled flowers are destroyed by the larval wasps that hatch from the eggs. These larvae later pupate.

Stigma
Short-styled flower
Ovipositor
Long-styled flower

2 Small simple female flowers line the inside of the fig. Some have long styles – stalks that connect the pollen-receptive stigmas to the flower's egg cells or ovules – and others short styles. Having unloaded her pollen on to the stigmas, the female wasp lays eggs near the flower ovules with her ovipositor. This is only long enough to reach the ovules of short-styled flowers. After laying her eggs the female wasp dies.

Female flower

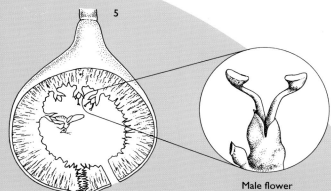

Male flower

The reproductive cycles of a fig tree and a species of fig wasp are intermeshed. The two organisms have coevolved to such an extent that neither can reproduce without the other.

5 Inseminated female wasps load up with pollen from the male flowers and leave the fig via the holes made by the males or through the opened ostiole.

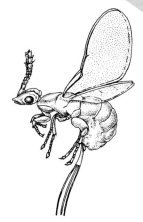

Female wasp

Male wasp

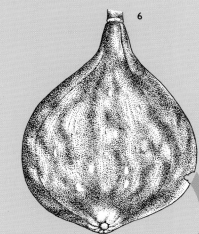

6 The winged females, inseminated and packed with pollen, are ready to start the cycle again. The fig they leave is filled with ripe seeds and now changes from a sour green fruit into a sweet purplish-brown one that attracts fruit eaters such as monkeys.

I An inseminated female fig wasp, carrying pollen in containers on her body, enters a young fig through the ostiole, a bract-covered hole at one end. In so doing her wings and the outer parts of her antennae are broken off.

determined way in which the female wasp mutilates herself in order to enter the fig fruit and lay her eggs — she often loses her wings and parts of her antennae in the process.

The inside of the fruit is lined by flowers with long and short styles. The female wasp inserts her ovipositor into a short style and lays her eggs in the ovary below — the short-styled flowers become "nurseries" for developing wasp larvae. Because the ovaries of the long-styled flowers are beyond the reach of the ovipositor, they are left undisturbed and able to produce useful fig seeds.

METABOLIC BENEFITS

Mutual interdependence can involve the coevolution of more than two species. The life strategy of termites, for example, revolves around symbiosis. Most termite species feed on cellulose-rich foodstuffs such as dry wood and leaves. Although the termites cannot digest the cellulose themselves, their intestines harbour a complex mixture of single-celled protozoan and bacterial species that can. Some of the bacteria even fix nitrogen from the air which the termites then use to make amino acids, the building blocks of proteins. The termites cannot live without their microbial partners, while the bacteria and protozoans can live only in the guts of termites. None of these species could survive without the others.

This species cooperation is just one of a highly varied range of symbioses featuring microorganisms, whose metabolic marvels make them valuable partners. Green plants, for instance, despite their apparent metabolic independence as a result of photosynthesis, often become involved in symbioses. Many plant families, particularly legumes such as peas and beans, associate with soil-dwelling, nitrogen-fixing bacteria in the genus *Rhizobium*.

In these symbioses, different strains of *Rhizobium* are specific for different legume species. The bacteria attach themselves to a plant's root hairs and are transferred to the cytoplasm of root cells where they form the root nodules characteristic of legumes. Bacteria in the

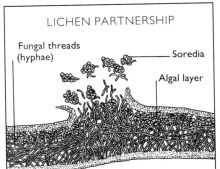

LICHEN PARTNERSHIP

Fungal threads
(hyphae)

Soredia

Algal layer

Multicoloured encrusting lichens cover
rock surfaces where few other
photosynthetic organisms could survive
(*left*). A lichen is a partnership of two
organisms. It is mainly composed of
fungal threads (hyphae) woven into a
dense feltwork. Near the upper surface
of this mesh is a layer of symbiotic algal
cells which can photosynthesize.

Lichen multiplies asexually when
small mixed clusters of hyphae and algal
cells known as soredia are dispersed
from the surface of the lichen (*above*).
Each soredium can grow into a new
lichen consortium.

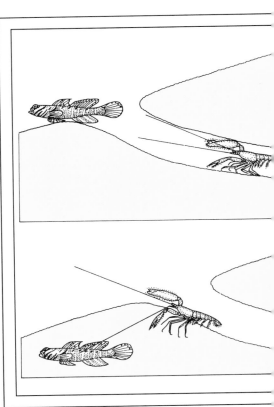

nodules fix nitrogen from the soil's air spaces and make it available to the plant for use in the manufacture of amino acids and nucleotide bases. This partnership means that legumes can be exceptionally productive and protein-rich, even in soils that lack nitrogen-containing salts such as nitrates.

WITH THE HELP OF A FUNGUS

In a partnership known as a mycorrhiza a plant root intermingles with the thread-like hyphae of a specific soil-dwelling fungus. The hyphae, anchored into cells within the plant's roots, spread out like a halo in the surrounding soil. Although the fungus gains some nutrients from the plant, the plant gains considerably more because the hyphae increase its capacity to absorb mineral salts from the soil.

Mycorrhizae become particularly crucial in conditions where essential minerals such as phosphates are a limiting resource – short supplies of phosphate restrict the amount of growth a plant can make. In these circumstances, a mycorrhizal fungal partner greatly improves the

fitness of some conifers, heathers and crop plants such as coffee.

THE FOOD OF ALGAE

Photosynthetic microbes, such as single-celled green algae and blue-green algae, feature in many symbioses because their metabolic products can be used as food by their partner organism. Some protozoans, sponges, flatworms and coelenterates benefit from the presence of algae in their tissues. When the animal moves into sunlight, the algae photosynthesize. In return for making carbohydrates freely available to the animal's tissues in this way, the algae gain protection, carbon dioxide and some minerals.

Such symbioses can have major ecological consequences. The production of coral reefs by coral-budding coelenterates, for instance, depends on the dinoflagellate algae of the genus *Symbiodinium* which live inside coral polyps. They are contained within the gut cells of the polyps and are thought originally to have been acquired as food items which were retained instead of being digested.

When the algae photosynthesize, the polyps produce calcium carbonate, which forms hard material or "rock"; when the algae are unable to photosynthesize, rock formation ceases. Such photosynthetic needs mean that coral reefs are restricted to shallow water where there is sufficient sunlight. And without these symbiotic partners the Great Barrier Reef would not exist.

Photosynthesizing algae team up with fungi to form lichens which inhabit and thrive in extremely cold, dry or nutrient-poor environments. The thousands of different "species" of lichen are not really species at all. Instead, they are consortia of two or sometimes three species – one fungus and one or two algal species – that form a structure consisting of fungal strands or hyphae enmeshing a population of algal cells. Some lichens have one green and one blue-green alga, the latter partner fulfilling a nitrogen-fixing role.

GAINING PROTECTION

As well as supplying food for each other, partners in a symbiosis can also give

MUTUAL BENEFITS

Animals can gain evolutionary advantages by teaming up and living together. Alphaeid shrimps live on the sandy seabed and can dig burrows for protection. But these shrimps are blind and are in danger from predators whenever they leave their burrows. Goby fish also need places to hide from enemies, but they cannot burrow. But if the fish shares a shelter with the shrimp, both are better off. The fish has protection while the shrimp has a sharp-eyed guard.

The shrimp appears to have evolved a system of communication with the fish and leaves its burrow only when the fish is at the entrance. When the shrimp does come out, it keeps one of its antennae touching the fish. If danger threatens, the shrimp responds to the fish's escape movements and both immediately return to the burrow. In observations of the behaviour of these partners, the shrimp did not react to danger if the fish was removed, suggesting that it is dependent on this "warning system".

Another aspect of the partnership is that the shrimp cleans the fish. It removes and eats external parasites and damaged surface tissue.

LIFE FITS THE WORLD

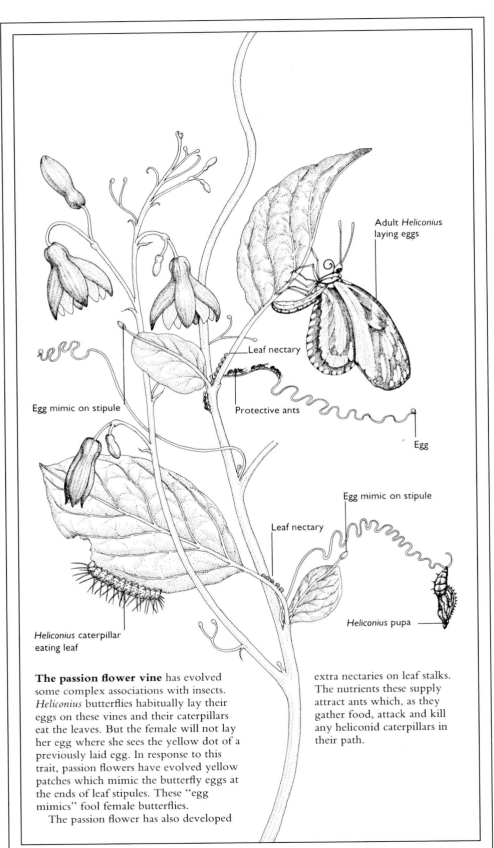

Adult *Heliconius* laying eggs

Leaf nectary

Egg mimic on stipule

Protective ants

Egg

Egg mimic on stipule

Leaf nectary

Heliconius pupa

Heliconius caterpillar eating leaf

The passion flower vine has evolved some complex associations with insects. *Heliconius* butterflies habitually lay their eggs on these vines and their caterpillars eat the leaves. But the female will not lay her egg where she sees the yellow dot of a previously laid egg. In response to this trait, passion flowers have evolved yellow patches which mimic the butterfly eggs at the ends of leaf stipules. These "egg mimics" fool female butterflies.

The passion flower has also developed extra nectaries on leaf stalks. The nutrients these supply attract ants which, as they gather food, attack and kill any heliconid caterpillars in their path.

mutual protection. For example, the sandy sea bottom environment of gobies and alphaeid shrimps offers little refuge from predators. The shrimp can dig burrows in which to hide but, being blind, is an easy target for fish when feeding outside the burrow. By contrast, the goby cannot burrow but has good eyesight. As a pair living together, they both gain – the goby a protective burrow, the shrimp an efficient sentry.

Protection also underpins some remarkable and complex associations between plants and insects, especially ants. Plants which adapt their structure to attract an ant colony inherit a well-armed social group whose members bring with them a fierce ability to defend their nests. Such a force is more than able to defend the plant from the enthusiastic attentions of foraging herbivores. The base of each spine on the branches of an ant acacia in Africa, for example, has evolved into an expanded black hollow where ants have taken up residence. When a browsing herbivore such as a gazelle starts to nibble the leaves at the end of a branch its snout is quickly covered by biting and stinging ant workers defending the plant.

In a parallel example, the west African tree *Barteria*, a member of the passion flower family, has a close relationship with large biting ants from the genus *Pachysima*. The ants live in the tree's hollow branches, gorging on the honeydew of scale insects which they "farm". Insect herbivores and leaf-eating monkeys attempting to eat the foliage of the tree are discouraged by the presence of hundreds of protective worker ants. Farmers who clear forest vegetation for arable land often leave *Barteria* trees untouched because of the swift insect retribution they know awaits them if these trees are felled.

In the rainforests of South America other members of the passion flower family take part in similarly complex partnerships. The successful and diverse passion flower vines suffer as a result of the voracious appetites of the caterpillars of butterflies belonging to the genus *Heliconius*. Each butterfly species lays its eggs on the leaves of particular vine

species. Not only are the larvae immune to the toxic chemicals produced by their host vine species but they are also able to defoliate and kill it. Such selection pressure has forced the vines to evolve measures for reducing larval attack (see box).

BROOD PARASITES

Host–parasite relationships are found throughout the animal and plant kingdoms. One sort of parasite – the brood parasite – does not use the host as a direct source of food. Instead, it induces the host species to feed and care for its young. The main evolutionary consequence of this devious strategy is that the brood parasite can produce far more offspring than it would do if it reproduced independently. However, the host species loses out reproductively – it is tricked into diverting food and energy to the success of another species rather than caring for its own young.

Although there are brood parasites among both fish and insects, the most well-known examples are found among birds. The common cuckoo, *Cuculus canorus*, epitomizes the brood parasite. When adult cuckoos migrate from Africa to Europe in the spring they return to a site close to where they were raised. In fact, the cuckoo species is split up into genetically differentiated subspecies called gentes – each gens is specifically adapted for parasitizing a particular host species. A cuckoo may belong to, say, the meadow pipit gens or the hedge sparrow gens. The main consequence of this differentiation lies in the fidelity of egg mimicry. Females in a particular gens can produce excellent mimics of the eggs of their host species.

A CUCKOO'S ADAPTATIONS

An individual female cuckoo defends an area of countryside containing many

A nestling cuckoo has pushed the much smaller natural offspring of its foster parents from the nest. They will die, leaving the host pair's reproductive success at zero for that season. The cuckoo chick, however, is likely to survive and breed.

nests of her host species which she observes regularly. After a male cuckoo has inseminated her she begins to lay eggs in the nests – up to 30 in a season but never more than one per nest. The cuckoo usually times her egg laying to coincide with the early stages of host egg production.

Host birds have developed a "desertion" defence against cuckoos. If a host bird notices that eggs have been disturbed it will leave that nest and start another with new eggs. To counteract this defence cuckoos lay their single egg very fast, completing the task in the short period when the host bird leaves its nest unguarded. Although most birds take

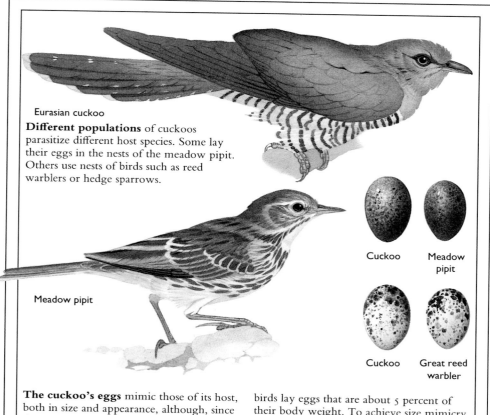

Eurasian cuckoo

Different populations of cuckoos parasitize different host species. Some lay their eggs in the nests of the meadow pipit. Others use nests of birds such as reed warblers or hedge sparrows.

Meadow pipit

Cuckoo   Meadow pipit

Cuckoo   Great reed warbler

The cuckoo's eggs mimic those of its host, both in size and appearance, although, since the cuckoo is larger than its hosts, its eggs might be expected to be much bigger. Most birds lay eggs that are about 5 percent of their body weight. To achieve size mimicry the cuckoo lays eggs that are only about 2.5 percent of her body weight.

The instinctive bond between parent and the young in its nest keeps the host bird caring for the cuckoo. This reed warbler (*left*) is supplying food to a fledgling cuckoo that is already much larger than its foster parent.

many minutes to lay an egg, a female cuckoo can deposit hers in 9 seconds – an obvious adaptation for brood parasitism.

Other egg-laying adaptations include the "egg delivery chute" – a protrusible cloaca which enables the female cuckoo to lay in a nest that she cannot physically reach. Her eggs also have particularly thick shells to help them withstand rapid laying. But perhaps the most astonishing feature of the eggs is the way they mimic the host's eggs so accurately. They match them in colour and patterning, and are also close to them in size (see box).

TAKING OVER THE NEST

After the female cuckoo lays an egg in a nest she almost always carries off a host

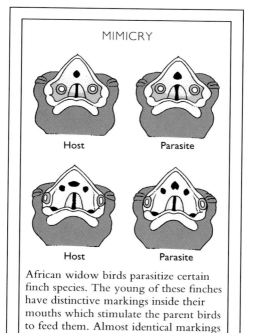

MIMICRY

Host   Parasite

Host   Parasite

African widow birds parasitize certain finch species. The young of these finches have distinctive markings inside their mouths which stimulate the parent birds to feed them. Almost identical markings have evolved in the widow bird young.

egg in her beak and destroys it, thereby ensuring that the nest retains the same egg number. Cuckoos evolved this behavioural adaptation in response to the basic counting ability which many birds probably developed as part of their defence against brood parasites.

As soon as the young cuckoo hatches it forces the other eggs and hatched young from the nest. All the host parents' efforts are channelled into feeding and protecting the young cuckoo even though it soon looks different and much bigger than their own young. Any tendency to evolve a rejection of this bizarre youngster would mean reducing the strength of the basic bonding between parent and young. Such a path might have dangerous consequences in the majority of cases where a parasitic youngster is not present in the nest.

As it grows up the young cuckoo becomes familiar with the habitat around the nest and the appearance, songs and calls of its host parents. The capacity to recognize these will aid the youngster when it returns to breed itself the following season. It will need to seek out the same type of countryside and nests of the same species of bird in order to achieve its own parasitic reproductive success.

VISUAL BRAND MARKS

In other similar pairings among birds defensive coevolutionary adaptations in the hosts have tried to outwit the brood parasite. The nestlings of the melba finch, which is a habitual host to the parasitic widow bird of Africa, have evolved visual brand marks inside their beaks (see box). Adult finches will only push food into gapes which have the correct markings.

It might be thought that this adaptation would completely rule out efficient parasitism – since if the widow bird's young lacked these patterns they would not be fed. In fact, selection pressures have driven a breathtaking piece of parasitic counter-adaptation – the widow bird's young have developed a similar set of gape markings which are good enough to convince adult finches to feed them. The coevolutionary arms race continues.

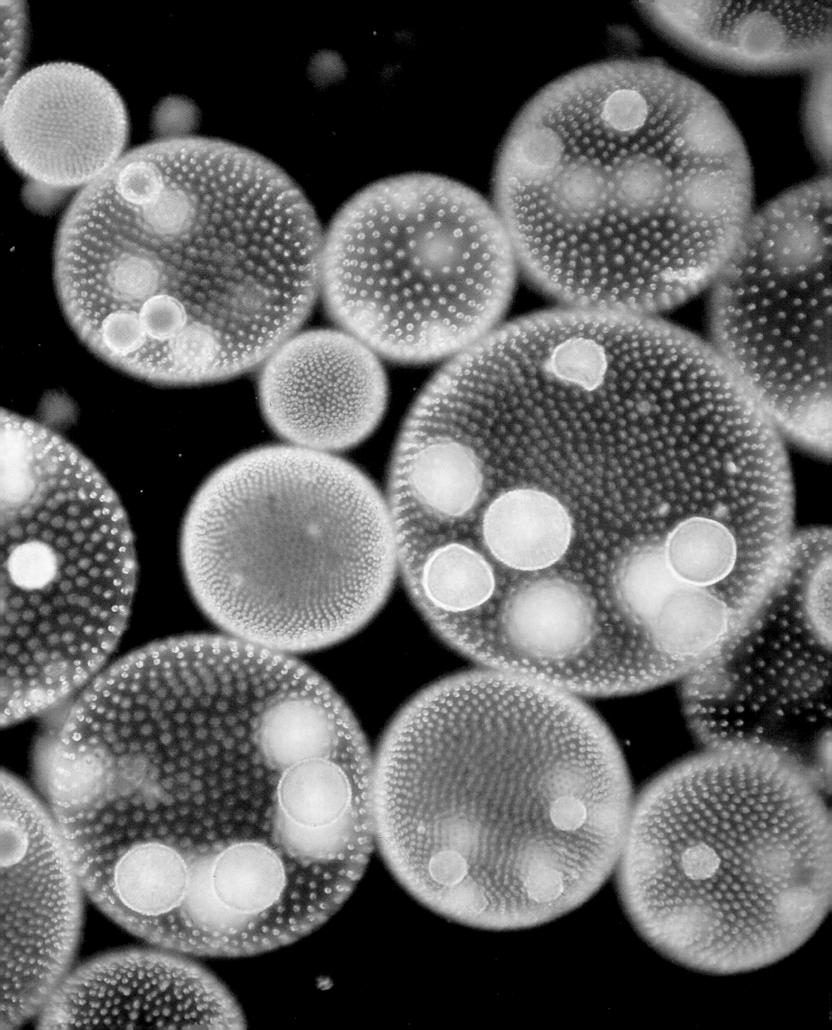

NEW PERSPECTIVES

Life changes inexorably with the passage of time. From moment to moment, natural selection sifts through the inherited variation in living things searching, as it were, for organisms best adapted to the prevailing conditions. Accumulation of selected adaptive traits in these organisms inevitably leads to the origin of new species.

These central ideas about evolution have survived intact from the days of Darwin and Wallace (pp. 10–17) in the 19th century. Yet, like all scientific theories, they have been continually renewed and refined. Throughout the first half of the 20th century, the advances made in understanding the mechanisms behind variation and selection – at the levels of both genetics and populations – would have amazed and delighted the early pioneers.

The advent of molecular genetics and molecular biology in the second half of the 20th century precipitated another time of renewal and refinement of the concepts at the heart of evolution. The modern interdisciplinary approach to the life sciences has viewed many aspects of evolution from surprising yet illuminating perspectives. Revelations about plate tectonics, for instance, offer an array of new evidence relating to both the distribution of species over the globe and the dynamics of macroevolution.

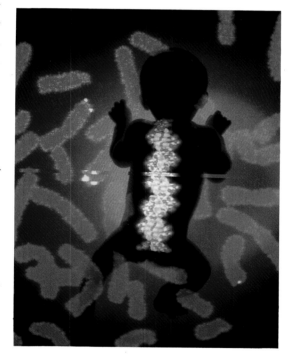

Molecular biological insights into the structure of genes have, arguably, transformed our world-view of evolution most. Incredibly detailed genetic information in the form of nucleotide base sequences has become accessible. Consequently, the meaning of evolutionary processes can be untangled, dissected and clarified with a precision undreamt of a few decades ago.

The British evolutionist Richard Dawkins has focused attention on the genes as the central players on the evolutionary stage. By suggesting that the bodies of animals are machines built by genes to promote their own survival and multiplication, Dawkins has raised fundamental questions about some features of Darwinian evolution as applied to life in its broadest, most universal sense.

Analysis of the genes of multicellular organisms, viruses and bacteria makes it possible to demonstrate the different evolutionary worlds in which these types of organism live. The complexity of eukaryote evolution, for example, is partly attributed to the emergence of developmental genes as controllers of cell differentiation in an embryo.

Viruses, viewed as almost naked genes, are close to Dawkins's concept of the ultimate organism. Bacteria, as typical prokaryotes, are revealed as quick-change artists that can transfer genes between each other in processes that are the closest that bacteria get to sexual reproduction.

The extreme flexibility of some parts of the bacterial genome may be the explanation for some surprising recent research. This suggests that some kind of "purposeful" evolution exists. Bacteria that lack a gene enabling them to digest a particular nutrient seem to acquire one when they are placed in that nutrient. And they do so more quickly than could be possible by random mutations. There are echoes of Lamarckism (pp. 11–12) in these as yet inexplicable findings.

With these molecular insights have come associated powers to modify genes and to engineer evolution itself, at least in the laboratory. As scientists dig deeper into the nature of life, they not only increase the total landscape of illuminated evolutionary fact but they are also forced to confront ethical issues that may yet decide the future of the planet.

CHANGE ON A LARGE SCALE

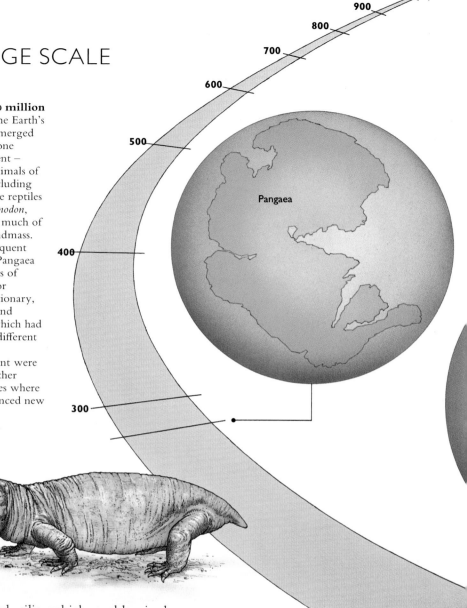

The identity of a species is never fixed but is subtly changing all the time. Its individual members, congregated into one or more populations, are part of a gene pool that can generate vital, small-scale adaptations in genotype and phenotype (pp. 66–67). The consensus among experts is that these small-scale, micro-evolutionary adaptations represent the raw material of evolution as a whole.

But how do large-scale, macroevolutionary changes, such as the creation of eukaryotes or the generation of a new phylum, take place? Are they simply the combination of many small-scale adaptations or are they partly the consequence of other contributory factors acting on populations in both space and time?

One important contributory factor on this macroevolutionary stage was endosymbiosis (pp. 28–29). This process, in which one prokaryote took up residence within another, almost certainly produced some of the key attributes of early eukaryotes. Through endosymbiosis, aerobic bacteria probably gave rise to mitochondria, while free-living cells similar to blue-green algae may have developed into photosynthesizing chloroplasts. The evolution of these cytoplasmic organelles was a vital macroevolutionary event that opened up new metabolic possibilities for aerobic and photosynthesizing eukaryotes.

American biologist Lynn Margulis and others committed to the importance of endosymbiosis include additional symbioses in their macroevolutionary scenario. For instance, they suggest that threadlike prokaryotes similar to spirochaete bacteria were incorporated into early eukaryotic cells and gave rise to the organelles called microtubules. A typical cell might then have played host to a pre-mitochondrion cell, a pre-chloroplast cell and a microtubule-containing prokaryote.

Microtubules are key components of the filamentous framework that gives eukaryotic cells their diverse and characteristic shapes. They also make up the centrioles, part of the spindle apparatus that divides up the sets of chromosomes at mitosis, and the inner scaffolding of

Around 250 million years ago the Earth's plates were merged together in one supercontinent – Pangaea. Animals of the time, including mammal-like reptiles such as *Dicynodon*, ranged over much of this huge landmass.

The subsequent breakup of Pangaea led to a series of large-scale, or macroevolutionary, pressures. Land organisms which had evolved on different parts of the supercontinent were taken into other climatic zones where they experienced new patterns of competition.

Dicynodon

flagella and cilia which enable single eukaryotic cells to swim.

If the cell partnership story is true it means that a human sperm swims with a flagellum whose structure is derived from an ancient prokaryote and whose power is generated in a mitochondrion, itself derived from another bacterium.

Constraints in development, which effectively block the elaboration of structure beyond a certain point, are also thought to have contributed to the macroevolution of living things. Moreover, such constraints seem to be essentially random.

Early four-footed vertebrates, for example, settled on the standard and characteristic five-toed (pentadactyl) limb after a period of less constrained experimentation with feet bearing different numbers of digits. A five-toed limb is probably no more or less efficient

than a six-toed foot, so the selection of five as the primitive condition in all major tetrapod groups is likely to have been random. Similar constraints apply to the fivefold symmetry of all echinoderm groups and the three-petalled flowers of all monocotyledonous flowering plants.

Continental drift provides another possible contributory factor to macroevolution. The slow ballet of the continents over the surface of the globe during the past 400 million years has coincided with the entire evolution of land-living animals and plants. As the tectonic plates that created the backdrop to this scene

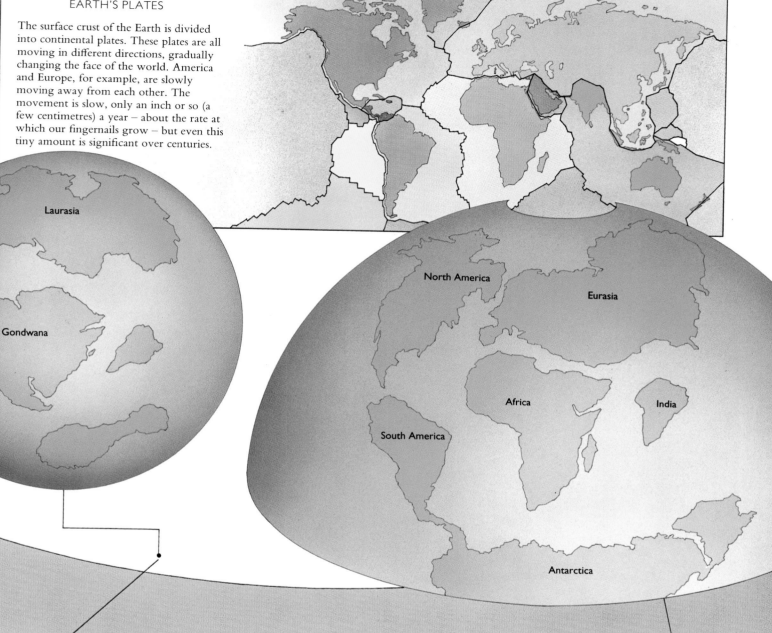

EARTH'S PLATES

The surface crust of the Earth is divided into continental plates. These plates are all moving in different directions, gradually changing the face of the world. America and Europe, for example, are slowly moving away from each other. The movement is slow, only an inch or so (a few centimetres) a year – about the rate at which our fingernails grow – but even this tiny amount is significant over centuries.

Laurasia

Gondwana

North America

Eurasia

Africa

India

South America

Antarctica

Million years ago

70

Protea

Plants of the Protea family (*left*) are found only in South America, Africa, southern Asia and Australasia. These landmasses made up the southern continent Gondwana. The ancestors of the Protea family probably evolved before Gondwana began to break up about 160 million years ago.

Horned dinosaurs, such as *Triceratops*, lived during the late Cretaceous in western North America. This family of dinosaurs has been found nowhere else in the world, which suggests that they evolved after North America had separated from other landmasses.

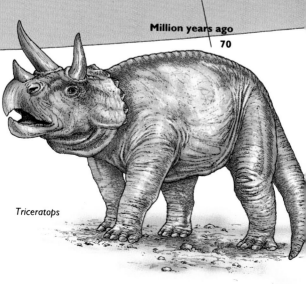

Triceratops

have split apart, collided and slid past one another, they have constantly rearranged the stage on which the drama of terrestrial evolution has been enacted.

Imagine a game of chess in which the pattern of black and white squares was always changing. Moves that obeyed one set of rules would become "illegal" as the board shifted. In the same way, inexorable continental changes over the past 400 million years have permitted some interactions, prevented others and generally shaped the competitive forces of macroevolutionary history.

Continental movements can influence evolutionary pressures by altering the physical environment. A tectonic plate,

for example, can move a continent from a tropical to a polar latitude. The concentration of landmasses around the high latitudes of the northern hemisphere in the last few million years has brought about a feedback loop of cooling influences in the area which, in turn, have created a large zone of ice and permafrost.

Land and sea barriers generated by continental drift have, by restricting movements, influenced zoogeographical distribution patterns on the face of the Earth. Organisms that arose and diversified on an ancient landmass, such as Gondwana, have been prevented by large sea barriers from colonizing other

landmasses. For example, plants in the Protea family are still today restricted to those landmasses, such as Africa, that are descended from Gondwana. The most crucial evolutionary effect of these patterns of life determined by continental drift is to do with competition. The life forms present or absent in a particular part of the world help to define the evolutionary fate of all the other organisms in the community.

One well-researched instance of a distribution defined by continental drift is the evolutionary status of the indigenous mammals of Australasia. Australia and the other islands that sit on the same tectonic plate possess a majestic diversity of mammalian types that are specialized and precisely adapted for the fullest possible range of niches on land.

This diversity arises almost entirely from an ancient subset of mammalian

Evolution does not happen in isolation. Within a habitat, what one species does affects another. This diagram illustrates such change, known as character displacement, through the feeding habits of two closely related species of saltmarsh snails (*Hydrobia*).

The snails feed on detritus and algae. Curve 1 plots the food resources available in the marsh, showing that most is of medium size. Curve 2 shows the pattern of food use by species A when living alone without competition from B. The curve mirrors the available resource – species A when living alone mostly takes medium-sized food items.

Species B may also live in this same habitat but takes proportionately more of larger size food items, as shown by the striped portion of the curve 3. Curve 4 shows the overall food use of species B when coexisting with A. In response to this competitive change in available resources, species A starts eating smaller food items. The result of competition is that A changes its behaviour.

Thus the presence of B has changed the food use behaviour of A. If A and B always shared a habitat this character displacement could become "fixed" as part of the behaviour of A controlled by its genes and passed down through the generations.

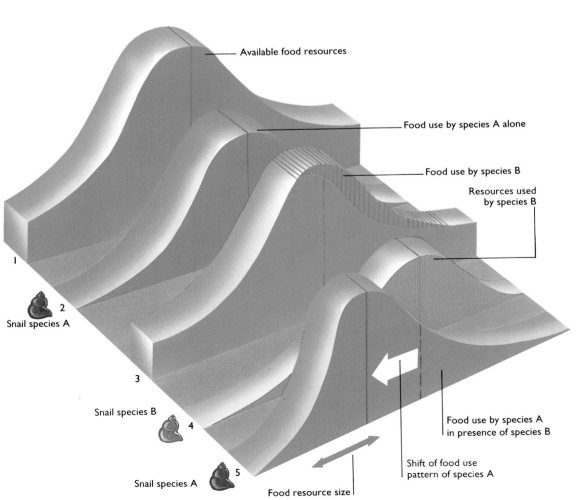

MADAGASCAR

Many families of mammals on the island of Madagascar, off the coast of east Africa, are endemic – they are found nowhere else – and are early offshoots of the groups to which they belong. The primates, so diverse in mainland Africa, are represented in Madagascar only by the lemurs, such as the ring-tailed lemur (*right*). Tenrecs, also a group found only on the island, are the sole representatives of the insectivore mammals.

For a long time it was thought that these ancient and endemic groups were living on the island when it drifted apart from Africa. Subsequent colonization by more advanced groups was then impossible and the endemic animals evolved into many species.

Modern evidence shows, however, that the story is not quite so simple. Madagascar is now thought to have finally split from Africa about 170 million years ago at a time when no placental mammals such as primates and insectivores existed. What may have happened is that after the separation of Madagascar the ancestors of the endemic species somehow got across from Africa. They may have moved across intervening islands or been carried by currents. Subsequent changes in the current flow or the disappearance of island stepping stones may have then prevented any later groups colonizing Madagascar.

stock – the marsupials. These pouched mammals, such as kangaroos, which do not mature their young in a uterus with a placenta, were an early line of mammalian evolution. Together with the egg-laying monotremes, for example the duck-billed platypus, they predated the placental mammals, but the latter now dominate natural terrestrial ecosystems worldwide – except in Australia.

When Gondwana split up into continental fragments, the one bearing Australasia probably moved apart from the rest before placental mammals first appeared on the macroevolutionary stage. Once they emerged into the spotlight, apparently initially in Asia, they colonized the remainder of the globe, eliminating in the process – presumably in competitive battles – almost all the monotreme and marsupial species that initially coexisted with them. A tiny remnant of this marsupial stock now survives in the Americas.

Placental mammals seem to be absent from the Australasian fossil record and the only living species, apart from some recent colonizing bats and rodents, are those that humans imported. Dingos were brought in about 8,000 years ago (by Aborigines from Asia), while sheep, cows and rabbits are more recent.

A particular pattern of continental movements in both space and time therefore had a profound effect on the fauna of Australasia, creating conditions in which marsupials could adapt and evolve for a very long period without competition from placentals. Some very recent fossil evidence from Australia – teeth from what could be a placental mammal – has raised the possibility that placentals and marsupials did interact competitively for a short time in that continent. If this did happen, the pouched mammals were ultimately victorious there.

Competition between similar organisms – mammal versus mammal, snail versus snail, weed versus weed – can also lead to macroevolutionary change. Such competition can reach a status quo if, when confronting each other in the same habitat, the species undergo so-called character displacement. By adapting their characteristics to slightly different niches, they grow apart yet become fitter occupants of the habitat they share.

Plant-herbivore, predator-prey and host-parasite linkages between species may have similar coevolutionary effects (pp. 158–67). For example, a brood parasitic bird and the host bird species that brings up its young embark upon a kind of escalating "arms race" that shapes the evolution of them both. It is a fascinating instance of the way in which one species becomes part of the adaptive landscape of another. Moreover, the influence the first species has on the second's macroevolutionary fate may be as powerful as, say, climatic change.

ARCHEOLOGY OF THE GENES

The discovery of the structure of DNA and the breaking of the genetic code opened the door to understanding the messages written in the genes of every living thing. Just as the Rosetta Stone unlocked the meaning of ancient Egyptian hieroglyphs, the techniques of molecular biologists allow them to read, analyse and compare the DNA from different organisms and thereby help to unravel the evolutionary relationships between species.

The first step in comparing the DNA of living things is to select for investigation a protein that almost all organisms possess. Such a protein is very likely to be one that first evolved long ago when all the organisms being compared had a common ancestor. All aerobic organisms, for instance, use enzymes called cytochrome oxidases in the manufacture of high-energy ATP molecules. These enzymes are primeval proteins that are present in all aerobic bacteria, fungi, protozoans, invertebrates, vertebrates and plants.

The nucleotide base sequence of the gene that codes for each of these enzymes has been slowly changing for hundreds of millions of years. By comparing the base sequence of the same gene in two different species – A and B – it is possible to tell the time that has elapsed since they evolved from a common ancestor.

In this sense, the gene becomes a molecular clock that started ticking when A and B separated from their common ancestor. If the base sequences are almost the same in both genes then the time elapsed is short; if they are substantially different, then the time is long. The greater the differences, the longer the time elapsed. The diagram on the right shows a comparison of nucleotide base sequences coding for the enzyme cytochrome C oxidase across a range of organisms.

The base sequences in different genes and different parts of single genes mutate at varying rates. It is as though each gene clock ran at its own particular local rate. This means that many different gene comparisons must be made in order to reach reliable conclusions about the patterns of evolutionary relationships.

These patterns can then help evolutionists establish meaningful family trees for the species concerned (pp. 176–77).

Remarkable new insights have been gained with this technique. Comparisons between the ribosomal RNAs of different prokaryotes, for instance, has revealed a hitherto unexpected schism. Long regarded as a single group, the prokaryotes are in fact two very separate kingdoms – eubacteria and archaebacteria (pp. 56–57) – with long histories of independence.

Since the 1970s, scientists have, with increasing regularity, discovered traces of genetic material in fossils, thereby paving the way for comparisons between fossil DNA and living DNA. Such comparisons became feasible with the development of a technique for amplifying tiny amounts of DNA. This PCR (polymerase chain reaction) technique turns DNA traces into amounts large enough for their base sequences to be read. This in turn has initiated the archeology of the gene.

At first, scientists focused on relatively recently dead material. Human DNA

Yeast, insects and humans all have an enzyme in their bodies called cytochrome C oxidase, but the genes that code for its production are not identical. A comparison of the differences in the nucleotide base sequence of the gene that codes for this enzyme reveals much about evolutionary relationships. The figures shown are the nucleotide base differences between humans and other organisms.

The pattern of these relationships is much as might be expected from indirect evidence such as comparisons of body structure. Base sequences from human and pig genes (both mammals) are more similar than those of a human and tuna (both vertebrates but one a mammal, one a fish). Such studies confirm that the lines of evolutionary reasoning from indirect evidence generally tally with those based on direct genetic comparison.

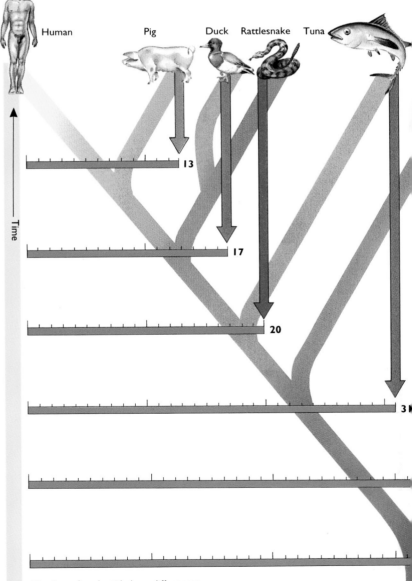

Number of nucleotide base differences

from Egyptian mummies a few thousand years old proved readable, as did the DNA of 40,000-year-old frozen mammoths. The DNA archeologists then leaped farther back in time by reading, for example, the DNA from insects fossilized in amber 25 to 40 million years ago.

DNA from magnolia trees was preserved when, some 20 million years ago, leaves were pressed in the cold, oxygen-free sediments at the bottom of a swamp in Idaho in the United States. The PCR technique has enabled scientists to extract fragments of a gene called *rbcL*, which is involved in photosynthesis. Base sequence comparison with the same gene in modern magnolias revealed that, out of a total of 820 bases, only 17 had changed in 20 million years.

Thus, a gene coding for a protein that helped trap the sunlight illuminating an Idaho swamp 20 million years ago has ended up on a laboratory bench in California. Like a molecular crystal ball this ancient gene helps to reveal the evolutionary history of one group of flowering plants.

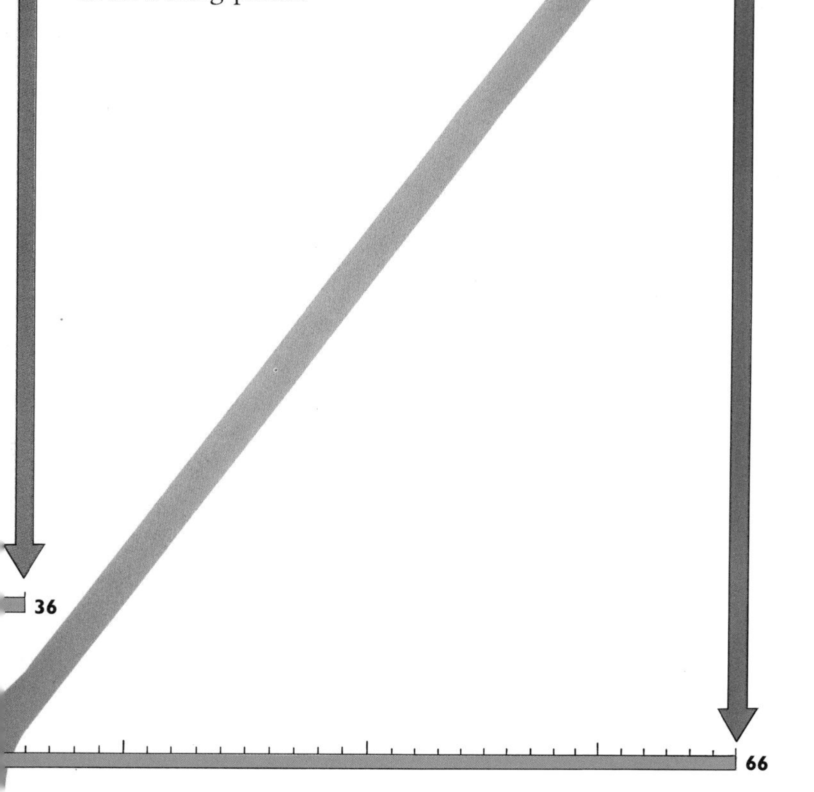

Moth

Yeast

36

66

THE QUAGGA REBORN

Hunted to extinction on the plains of South Africa, the last known quagga died in 1883. A zebralike animal, it had a brownish coat with distinct stripes only on its forequarters. But the quagga could live again. DNA from preserved skin has revealed that the quagga was a subspecies of the zebra, not a separate species as once thought. Scientists believe that quagga genes survive in the zebra population and by generations of selective breeding of brownish zebras with poorly striped coats they hope to produce quaggalike animals.

THE EVE CONTROVERSY

The publication in 1987 of an analysis of the mitochondrial DNA of modern humans from different parts of the world led to what has been termed the Eve hypothesis. Computer-based comparisons of the nucleotide base sequence of this genetic material suggested that there was a significant difference between most African samples and those of all other indigenous humans from Europe, Asia and Australasia.

These patterns were interpreted as meaning that modern *Homo sapiens* evolved from more ancient forms of the species in Africa. The research was also held to mean that a small colonizing group of these new humans had moved out of Africa and their descendants then diversified into new racial types as they spread into different habitats. This "break out" from Africa was thought to have taken place between 100,000 and 200,000 years ago.

Although many scientists still believe that modern humans began in Africa, there are doubts about the computer analysis which led to the Eve hypothesis. More recent work suggests that the original mitochondrial DNA data could be used to produce many evolutionary sequences, not all of which would indicate the existence of a small founder group of African origin.

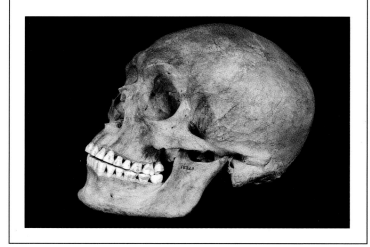

EVOLUTION AND CLASSIFICATION

If a wolf, a lion and a tiger are classified by phenetic and then by phylogenetic systems the same pattern is revealed. In their evolutionary histories, the lion and tiger are more closely related to each other than either is to a wolf. Structural (phenetic) analysis gives the same result. The mauve diagram below shows the actual evolutionary history of these animals, related to time, and their degree of structural similarity.

Phenetic classifications of groups of animals are determined entirely by similarities and differences in structure. The evolutionary history of the groups is not taken into account.

Phylogenetic classifications are defined by known or deduced facts about the evolutionary relationships of the animals concerned.

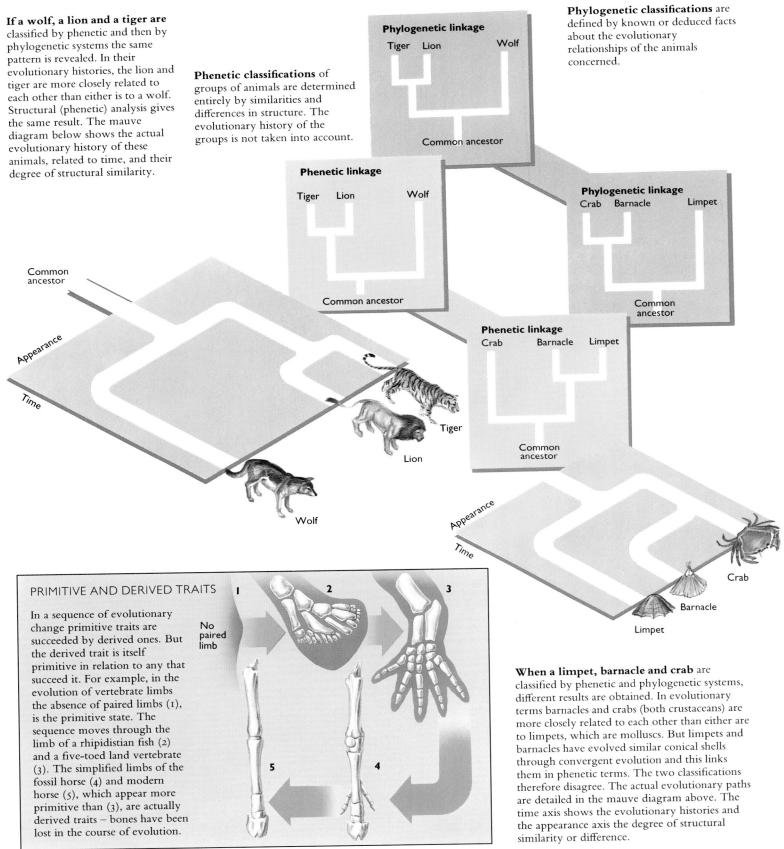

Phylogenetic linkage

Tiger Lion Wolf

Common ancestor

Phenetic linkage

Tiger Lion Wolf

Common ancestor

Phylogenetic linkage

Crab Barnacle Limpet

Common ancestor

Phenetic linkage

Crab Barnacle Limpet

Common ancestor

Common ancestor

Appearance

Time

Tiger

Lion

Wolf

Appearance

Time

Crab

Barnacle

Limpet

PRIMITIVE AND DERIVED TRAITS

In a sequence of evolutionary change primitive traits are succeeded by derived ones. But the derived trait is itself primitive in relation to any that succeed it. For example, in the evolution of vertebrate limbs the absence of paired limbs (1), is the primitive state. The sequence moves through the limb of a rhipidistian fish (2) and a five-toed land vertebrate (3). The simplified limbs of the fossil horse (4) and modern horse (5), which appear more primitive than (3), are actually derived traits – bones have been lost in the course of evolution.

No paired limb

1 2 3

5 4

When a limpet, barnacle and crab are classified by phenetic and phylogenetic systems, different results are obtained. In evolutionary terms barnacles and crabs (both crustaceans) are more closely related to each other than either are to limpets, which are molluscs. But limpets and barnacles have evolved similar conical shells through convergent evolution and this links them in phenetic terms. The two classifications therefore disagree. The actual evolutionary paths are detailed in the mauve diagram above. The time axis shows the evolutionary histories and the appearance axis the degree of structural similarity or difference.

The urge to classify living things is shared by scientists and non-scientists alike. Everyone can accept without too much prompting that, despite their obvious differences, eagles, ostriches and hummingbirds are all birds. Yet the major groups, or taxa, of living things are not simply a tidy way of organizing the millions of species that exist on Earth. The categories of life and their relationships in a connected hierarchy can be construed as a pared-down diagram of the very process of evolution.

Modern taxonomists wrestle constantly with the best way of constructing a classification system that is meaningful and not arbitrary. What is it that makes, for example, the phylum Mollusca seem more "real" and less arbitrary as a group than, say, all those animals with red bodies? It is because the snails, slugs, squids, bivalve shellfish and other groups within the Mollusca share a common evolutionary history.

A workable classification system must comprise a hierarchy in which groups are contained completely within larger composite groups with no overlap. The human species, for example, belongs to the genus *Homo*, which is a member of the primate order, which is in the class Mammalia and so on.

Two classification systems – the phenetic and the phylogenetic – are feasible. In the former, each group in the hierarchy is determined entirely by similarities of structure. In the latter, the groupings are based on the pattern of their evolutionary history.

The outcomes of classifying different trios of species according to the two systems are shown in the diagrams. In the case of the lion, tiger and wolf the outcomes are the same, but in the other, where convergent evolution has occurred, they are different. Convergence occurs when members of phylogenetically distant groups assume similar phenotypes in response to similar selection pressures. A shark and a dolphin, for example, have both evolved a streamlined body shape despite the fact that one is a fish, the other a mammal. Similarly, convergence in body form between a barnacle and a limpet hides the phylogenetic distance that separates them.

Many advocates of the phenetic approach employ numerical taxonomy to organize their hierarchies. By considering a sufficient number of different traits, or characteristics, they hope to arrive at an averaged degree of similarity or difference between species. Groupings of "close" species should then almost choose themselves. Unfortunately, there is no non-arbitrary way of averaging similarities, so ultimately this technique is itself arbitrary.

Taxonomists increasingly use the phylogenetic system called cladistics, devised in the 1950s by the German entomologist Willi Hennig. Cladistics defines taxa according to the recent common ancestors of species. If a species has a detailed fossil history its cladistic history – a kind of family tree – can be precisely established. Without the benefit of fossils, taxonomists must employ more indirect means, such as the concept of primitive and derived traits (see box). The traits of a species alter through time by a sequence of mutational changes and selection. Traits from an early phase are called primitive; those from a later phase are derived.

The central principle of Hennig's system is that "shared derived traits show common ancestry while shared primitive traits do not". No actual classification is based on a single shared derived trait. In practice, reliable classifications are produced when a range of comparisons, such as outgroup comparison (see diagram), point to the same conclusion.

The vast majority of taxonomists use a form of phylogenetic system that is wedded to the principle of evolutionary change. Indeed, the system presupposes evolution as the ultimate yardstick for its "naturalness". Once the actual pattern of ancestry of a group of species is known, the group can be classified absolutely. Far from being a dusty set of labels for organizing museum specimens, this classification system mirrors the evolutionary past of the groups of creatures it tries to describe.

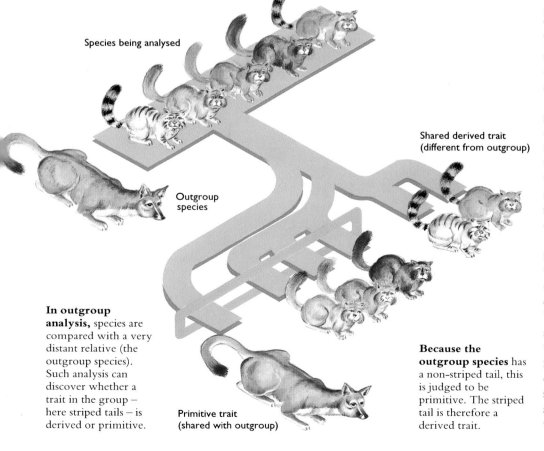

Species being analysed

Shared derived trait
(different from outgroup)

Outgroup
species

**In outgroup
analysis,** species are
compared with a very
distant relative (the
outgroup species).
Such analysis can
discover whether a
trait in the group –
here striped tails – is
derived or primitive.

Primitive trait
(shared with outgroup)

**Because the
outgroup species** has
a non-striped tail, this
is judged to be
primitive. The striped
tail is therefore a
derived trait.

THE TEMPO OF EVOLUTION

The fossil record harbours a wealth of information about the evolution of species and the history of life on Earth. But because fossils are mere shadows of the creatures that generated them, and because the fossil record is frustratingly incomplete, theories about speciation and the tempo of evolution are fraught with uncertainty.

A new theory about macroevolutionary change, proposed by Niles Eldredge and Stephen J. Gould in 1972, forced scientists to look again at their interpretations of the fossil record. The two US paleontologists challenged the orthodox theory that all evolutionary change was gradual by proposing that the predominant tempo of evolution was "punctuated equilibrium".

Looking again at previously described fossil sequences Eldredge and Gould found many examples in which, over a long period of time, species appeared to be unchanged in body structure. This is the static, or "equilibrium", part of their model. They then realized that there were times when the species would be abruptly, almost instantaneously, replaced by a new form. This species replacement represents the "punctuation" part of the model. Thereafter, the new form would enter a period of stability.

The central inference of this model was that most macroevolutionary change is not the result of slow and steady adaptive mutations as proposed by the orthodox, gradualist model. Rather, such change was predominantly the result of the abrupt creation of new species after long periods of inactivity or evolutionary stasis.

Moreover, evidence from the fossil record seemed to undermine the genetically orientated, mutationally based model of evolution. Eldredge and Gould argued that most large-scale patterns in evolution cannot be adequately explained by the accumulation of the types of changes seen in living populations through time (pp. 74–75). In other words, they were calling into question the accepted belief that microevolutionary change blends seamlessly into macroevolutionary change.

One study that appears to support the punctuationist model was carried out on a complete and apparently continuous fossil record of freshwater snails and bivalves from the Lake Turkana basin in east Africa. The sequence represented the last 4.5 million years of evolution of a series of lake-dwelling mollusc species. For long periods in this sequence, the snails and bivalves seem to have settled into an equilibrium in which they changed hardly at all.

At two key moments in the sequence, something remarkable appeared to have taken place in the now fossilized community. Almost every species present had been abruptly replaced by another form – another species of the same basic type. Apparently, each species had simultaneously and at a stroke altered into a new one. The whole dynamic of evolutionary change appears to have been concentrated into sudden bursts of creation rather than spread out through time. In other words, this data seems to suggest that the rate of evolution is far from constant and can vary from an almost imperceptible rate to one so fast that it appears from the fossil record to be instantaneous.

Other equally continuous fossil records, however, support a different interpretation. One study has examined the incredibly detailed fossil history of a microscopic protozoan in a group called the foraminifera, which are hard-shelled, single-celled organisms that float in the plankton. Their calcareous shells sink to the seabed where they become an important constituent of some marine sedimentary rocks.

The fossil sequence of these foraminifera stretched from just over 10 million years ago until about the present day. For the first 4.5 million years the shells varied slightly in size and thickness but with no

There are two contrasting ideas about the pace of evolution. One holds that evolution is gradual – a progression of slow steady change. The other sees it as long periods of inactivity, interspersed with bouts of rapid change. These two diagrams show the essential differences between the two concepts. In each, the progression of evolutionary change is visualized as a rolling wheel. The track left behind is the evolutionary history of that wheel.

The lower wheel, illustrating the gradualist theory, is made up of many tiny facets. Each of these represents a single mutation, which in turn produces an adaptation. This wheel rolls gradually and smoothly by the accumulation of these small mutational changes.

The upper wheel has a few large facets. As it rolls forward it stays for a long period on each facet – a period of evolutionary stasis or equilibrium. The wheel then abruptly turns to the next facet. This represents a "punctuation event" – the rapid flip from one species to another with many differences evident between the two forms.

Punctuated equilibrium

Gradualist evolution

obvious directional trend. It was as though slow random changes were taking place in a single species. About halfway through the sequence, some 5.5 million years ago, the shape of the shells changed smoothly. Small and thin at first, the shells gradually became much larger and thicker. Over a period of about 500,000 years, a new species had been created and has retained much the same form to the present day.

This example alone (and there are many others) shows that the punctuated equilibrium model does not always fit the detailed fossil record. The foraminiferan fossil sequence resembles the type of pattern expected from a species which adapted smoothly, over half a million years, to a change in its selectional environment and, in the process, produced an organism different enough to be called a new species.

The conflicting conclusions from the

Lava mounds dotting the landscape in the Lake Turkana region of Kenya in east Africa hint at the long volcanic history of the area. Fossils embedded in layers of volcanic sedimentary rock have provided important evidence for the punctuated equilibrium theory of evolutionary change.

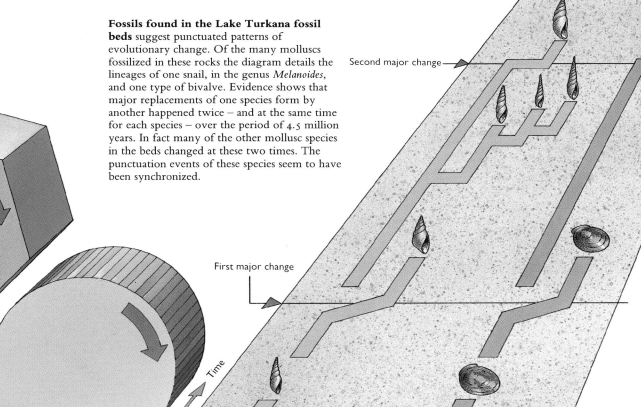

Fossils found in the Lake Turkana fossil beds suggest punctuated patterns of evolutionary change. Of the many molluscs fossilized in these rocks the diagram details the lineages of one snail, in the genus *Melanoides*, and one type of bivalve. Evidence shows that major replacements of one species form by another happened twice – and at the same time for each species – over the period of 4.5 million years. In fact many of the other mollusc species in the beds changed at these two times. The punctuation events of these species seem to have been synchronized.

Second major change

First major change

Time

Snail sequence

Bivalve sequence

Structural change

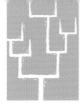

Turkana mollusc and foraminiferan fossil sequences make it difficult to judge, on paleontological interpretations alone, whether the gradualist or the punctuated equilibrium model is uniquely correct. Is there an unbridgeable gap, therefore, between those fossil sequences that demonstrate punctuated equilibrium and those that clearly indicate that populations change slowly through time?

A closer look at the Lake Turkana molluscs reveals that the evidence for punctuated equilibrium is not as watertight as was first thought. The long periods of stasis, in which about 10 species of gastropods and bivalves show little sign of structural change, do not necessarily require a special unorthodox explanation. Constant environmental conditions can act to stabilize a particular variant of a species. Such stabilizing selection tends to perpetuate the mean or average features of the species.

This stability does not prevent a species from generating new genetic variants; it simply means that in a specific set of environmental and competitive conditions such new variants are at a disadvantage and do not spread in the population. In this way, gradualist evolutionary ideas can easily explain long sequences of fossils which do not appear to be changing markedly.

With regard to the punctuations – the abrupt species replacements – that seemed to occur amongst the Lake Turkana shells, what did the fossil collectors actually observe? Over a tiny thickness of the total height of sedimentary rock sampled, a shell of one type was replaced – without the presence of intermediate forms – by a different type. It was the evidence of the fossils from this small thickness of rock that was interpreted as an abrupt species change.

A translation of rock thickness into real time units, however, suggests that the period of "abrupt" change could be between 5,000 and 50,000 years. The molluscs concerned would have been able to produce at least one new generation per year in tropical conditions, probably more. So the changes might have taken place over 5,000 generations at least – a perfectly adequate period of

Fossil evidence can be misleading. This hypothetical, yet perfectly possible, sequence illustrates how fossils may appear to support the punctuated equilibrium theory. The numbered stages on the time axis represent geological periods from the earliest (1) to the most recent (10). Ten stages in the evolution of a mollusc, corresponding to the ten layers of rock of different ages, are shown on the right-hand side. Also shown is the changing rate of sedimentation that leads to rock being laid down and allows fossils to be formed. The rate is high in periods 1 to 3 and 9 to 10 but almost non-existent in periods 4 to 8.

The actual evolution of the mollusc (shells 1 to 10) is gradual. But because little or no rock is formed in periods 4 to 7 there are no fossils from this period. Thus fossil evidence shows an apparent abrupt change between stages 3 and 9, appearing to support the punctuated equilibrium theory. If fossils for stages 4 to 8 were available, the real sequence would be seen to be gradual.

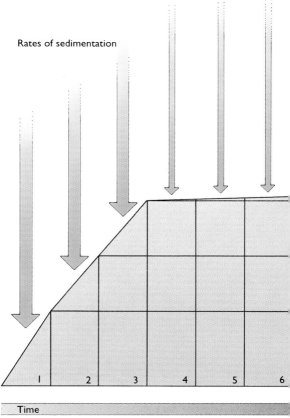

Rates of sedimentation

Time

time for normal adaptive changes to accumulate and produce a characteristic new phenotype.

Another feature of the Lake Turkana sequence is the way the species replacements were synchronized among almost all of the molluscs present at both moments of "punctuational" change. As a piece of circumstantial evidence this, by itself, points to an external, environmentally mediated influence. In fact, analysis of the Lake Turkana sediments demonstrates that at both "punctuations" the level of water in the lake rose or fell dramatically. During the long periods of equilibrium, the lake level experienced only minor fluctuations.

This correlation between a major environmental alteration and the relatively rapid creation of new species is almost certainly the true story behind the macroevolutionary change observed in the fossil sequence. A mollusc community, living for long periods in a relatively stable, predictable and unchanging environment within the lake, might well under stabilizing selection produce equally long fossil sequences of unchanging shells. Intervals of relatively rapid

DISTRIBUTION CHANGES

Variation in geographical range can cause what appears to be abrupt change. Ostracoderms were an ancient group of marine jawless fish that lived in different areas of what is now Europe. In this sequence, colour indicates the gradual evolutionary change in structure of these vertebrates as shown by fossils found in the Baltic area. This was near the centre of the distribution range of these ostracoderm fish, and here fossils have been found showing all stages of their evolution. (The Baltic area is represented by the central column on the diagram).

The sequence of fossils to the right of the central column shows apparent sudden changes between forms. These fossils came from what is now Scotland, an area only colonized by these creatures during occasional expansion in their range. These fossils give an erroneous impression of punctuated evolutionary change simply because of where they were found.

Total evolutionary sequence seen in the Baltic

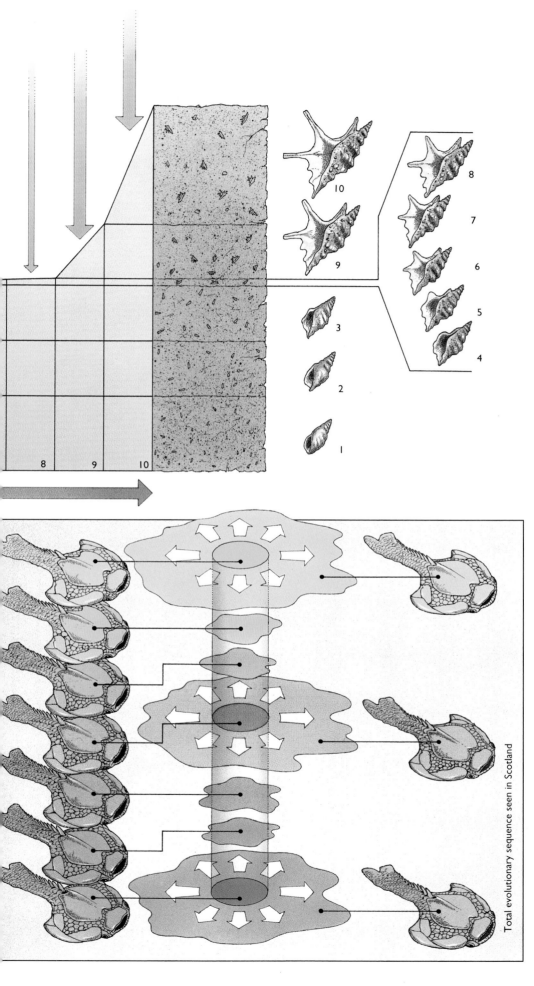

Total evolutionary sequence seen in Scotland

changes (albeit over thousands of years) in lake level would impose a new selectional pressure and drive morphological changes, such as alterations in shell structure, in the species present.

The case for punctuated equilibrium falters still further because the types of shell shape changes seen in the Lake Turkana sequence might not be species alterations after all. Present-day freshwater mollusc species show remarkable plasticity of shell form when they grow in various water conditions. Differences in water hardness, bottom substrates and other environmental influences can trigger phenotypic variability within the populations of a single species, such as Britain's common pond snail, *Lymnaea peregra*, with the result that shell form can vary widely.

The possibility that the abrupt "punctuations" of the Lake Turkana sequence were really displays of mollusc variability highlights the difficulty of reconstructing the dynamic population biology of a species from its fossils — except perhaps for those instances in which the fossils contain traces of DNA (pp. 174–75).

Although its most expansive claims are almost certainly untenable, the punctuated equilibrium model has been a most fruitful one for the scientific community. Eldredge and Gould have focused attention on the changing rates of evolution and forced paleontologists to re-evaluate the way they look at fossils as though they were surrogate live organisms.

Even the most precisely preserved fossil can only give a glimpse of the reality of the living organism. The fossil provides a partial description of the structural characteristics of the creature. That description, though, is almost always too imprecise to allow us to pontificate about the details of the boundaries between species and speciation events that took place millions of years in the past. When it is a commonplace that we are unable to tell the difference between male and female fossils, how can we expect to extract unambiguous information about the tempo of speciation from such data?

EXTINCTION AND EVOLUTION

The demise of whole species of living things played a vital role in the argument that finally won the 19th-century debate about the existence of evolution. Indisputable proof from the fossil record that just one species of organism had become extinct was enough to show that life changes through time. Once this fact was accepted evolutionists turned their attention away from extinctions as a biological phenomenon and concentrated much of their efforts on understanding how new species were produced.

The second half of the 20th century has witnessed a resurgence of interest in biological extinctions. First, ecologists have realized that the extinction of a species – whether it be animal, plant or microorganism – can have profound consequences on the populations of other species in its community. Second, mass extinctions of the past have captured the imagination of scientists and non-scientists alike. Third, there is a growing realization that human beings are responsible for the present, and possibly the most profound, mass extinction of species since life on Earth began.

Since species of organisms appear and disappear in the course of evolution, it is obvious that extinctions are a perfectly natural, all-pervasive characteristic of life on Earth. New species arise – usually when an already existing species splits in two – and disappear for ever with the same inevitability. Billions more species than those that live today existed in the past and have become extinct. The species present on our planet, which may number as many as 20 million, are poised between species birth and species death. As they become extinct, new species will appear to replace them.

The extinction of a species does not usually involve the sudden death of all its individual members. Rather, it is a function of the dynamics between rates of birth and death. Species will persist when their overall birth rate equals or exceeds their death rate. But if the latter exceeds the birth rate for a long enough period, replacement of one generation by the next ceases to occur. If no new factor intervenes then the species will eventually become extinct.

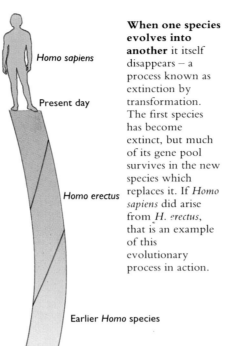

When one species evolves into another it itself disappears – a process known as extinction by transformation. The first species has become extinct, but much of its gene pool survives in the new species which replaces it. If *Homo sapiens* did arise from *H. erectus*, that is an example of this evolutionary process in action.

A whole group of species may become extinct, bringing a complete end to that evolutionary line. This happened to trilobites, a group of arthropods, at the end of the Permian about 245 million years ago. No trilobite genes survive in any existing arthropods because the evolutionary history of the trilobites has been separate from that of all other arthropods for so long – at least since Cambrian times.

Ultimately, a species extinction involves the loss of a gene pool. With every death of a species member the gene pool is depleted, but it is refilled with every birth. When refilling cannot replace depletion, the pool dries up and the species dies out. That particular collection of genes ceases to exist.

An accumulation of species extinctions can result in the disappearance of higher and higher taxonomic groups. With the demise of the last dinosaur species at the end of the Cretaceous, the whole dinosaur subgroup of the class Reptilia became extinct.

The disappearance of the dinosaurs from the face of the Earth was part of a mass extinction event when large numbers of species from a wide range of taxonomic groupings became extinct at about the same time (pp. 186–87). To identify such an event paleontologists

In an ice age polar and mountain range ice sheets are extended over normally temperate areas. For species which cannot migrate or adapt to the new cold conditions with no plant life, the extension of the ice brings severe risk of extinction.

must distinguish relatively sudden increases in the extinction rate over and above the background, or normal, rate of extinction. While paleontologists dispute what this background rate might be, they all agree that catastrophic numbers of species were lost after the Ordovician and Devonian periods, at the boundary between the Permian and Triassic and at the end of the Cretaceous period, 65 million years ago.

What caused the loss of 96 percent of all the species living during the Permian, or some 75 percent of those existing at the Cretaceous/Tertiary boundary? Most theories blame major climatic or environmental changes, such as those caused by continental drift, periods of glaciation that resulted in enormous changes in sea level, impacts of large objects from outer space or huge increases in volcanic activity.

The New Zealand kakapo, the world's largest parrot, is a flightless bird and feeds and nests on the ground. It is now nearly extinct for several reasons. Its flightlessness makes it vulnerable to introduced predators such as rats. It only breeds every four or five years so its rate of reproduction is slow. And since the female cares for her young alone, she has to leave them at night vulnerable to predators while she finds food.

Sea levels have not always remained the same through the history of the Earth. Fossil finds of marine creatures in the interior of continents indicate that sea levels were much higher at times and water extended farther into continents. There were also periods when sea levels fell. Both changes could cause extinctions.

The impact of large bodies from outer space may have caused extinctions on Earth. Such impacts could cause a nuclear winter with huge quanties of dust filling the air and smoke from fires blocking out the sun. Photosynthesis, the basis of life on Earth, would have been virtually impossible.

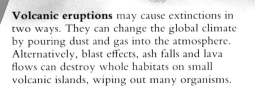

Volcanic eruptions may cause extinctions in two ways. They can change the global climate by pouring dust and gas into the atmosphere. Alternatively, blast effects, ash falls and lava flows can destroy whole habitats on small volcanic islands, wiping out many organisms.

MASS EXTINCTIONS

At the end of the Ordovician, about 440 million years ago, 85 percent of all species on Earth disappeared in what was the first major mass extinction. These species included many trilobites, brachiopods and nautiloids. Around 60 percent of all genera were lost at this time.

Numbers of species climbed following the Ordovician extinction period but, toward the end of the Devonian, about 380 million years ago, fell again. About 82 percent of species, 55 percent of genera, became extinct. These included many coral reef invertebrates and primitive fish species.

700

600

500

400

End of Ordovician mass extinction

End of Devonian mass extinction

End of Permian mass extinction

From among the many previous theories, a consensus view is emerging on one of the mass extinctions – that which occurred at the end of the Cretaceous. More and more evidence points to the impact of an extraterrestrial object, probably an asteroid or meteorite, as the cause of the dramatic changes of 65 million years ago (pp. 186–87).

This interpretation gives an important insight into the extraordinary unpredictability of mass extinctions. A plummeting asteroid that can change the climate of the whole planet is a massive chance element to be placed in the scales of evolution. It can conceivably wipe out an entire category of organisms, changing life on Earth at a stroke.

Such an alteration could upset the competitive balance between different groups of organisms, tilting it in favour of a previously sub-dominant group. In other words, a major change in the evolutionary fortunes of a group could be due to sheer unpredictable chance rather than inbuilt selective superiority.

This "lottery of life" scenario can be tested by careful analysis of the fossil sequences before, during and after the transition brought about by a mass extinction. Imagine a setting in which the species richness of two competing groups is relatively unchanging. Then a mass extinction transition occurs, causing one group to disappear or be greatly reduced while the other increases in species richness.

The lampshells (brachiopods) and the bivalve molluscs from the Permian/Triassic mass extinction period have been the subject of an analysis of precisely this sort. Although unrelated, both these animal groups contain two-shelled, filter-feeding, aquatic species that compete when they exist in the same habitat. Before the mass extinction the lampshells were more abundant but were slowly declining in species numbers whereas the bivalves were slowly increasing. At the end of the Permian, the lampshells suffered a huge decline in species richness while the bivalves flourished, with a rapid and sustained increase in species numbers.

This pattern of change suggests that, in this instance at least, multiple evolutionary mechanisms were at work. The fact that lampshell species numbers were declining before the transition suggests that the bivalves had some intrinsic competitive edge over them and were already expanding at their expense. What was probably a chance decimation of lampshells in the mass extinction seems to have accelerated the increase in bivalve species richness.

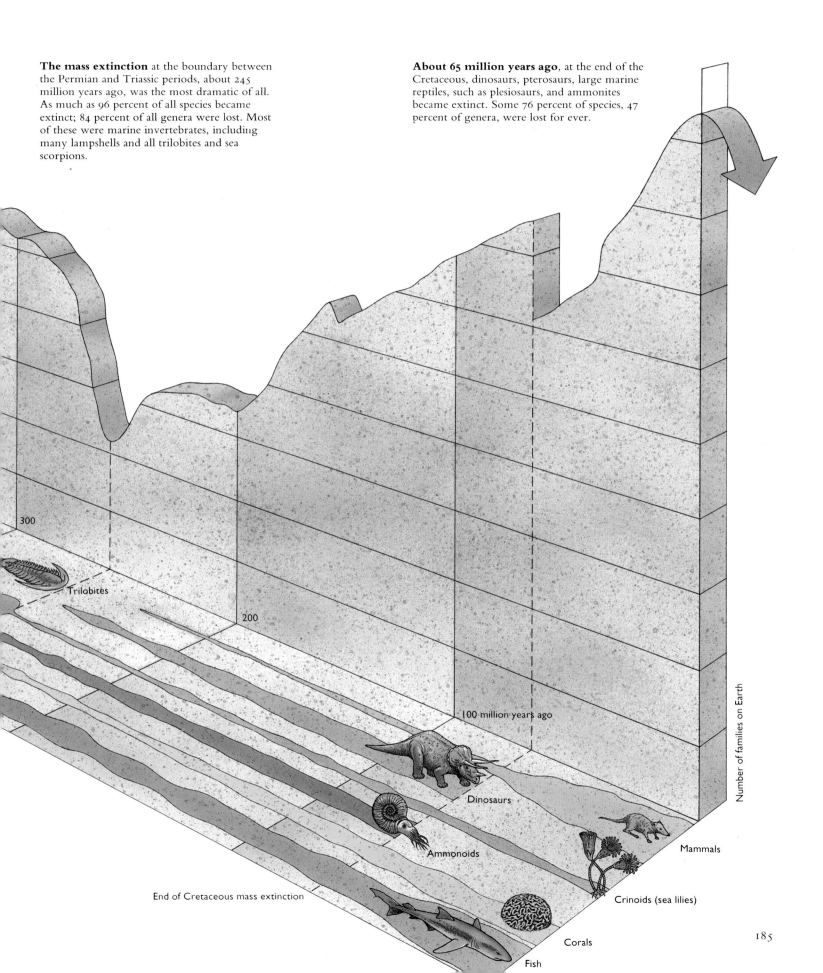

The mass extinction at the boundary between the Permian and Triassic periods, about 245 million years ago, was the most dramatic of all. As much as 96 percent of all species became extinct; 84 percent of all genera were lost. Most of these were marine invertebrates, including many lampshells and all trilobites and sea scorpions.

About 65 million years ago, at the end of the Cretaceous, dinosaurs, pterosaurs, large marine reptiles, such as plesiosaurs, and ammonites became extinct. Some 76 percent of species, 47 percent of genera, were lost for ever.

300

Trilobites

200

100 million years ago

Dinosaurs

Ammonoids

Mammals

Crinoids (sea lilies)

End of Cretaceous mass extinction

Corals

Fish

Number of families on Earth

THE END OF THE DINOSAURS

Around 65 million years ago, a catastrophic event killed off three quarters of the life forms on our planet. Every belemnite and ammonite mollusc, every flying reptile and every large marine reptile in the plesiosaur group became extinct. Most dramatically of all, the dinosaurs – a group of reptiles at the apogee of their species diversity – were completely wiped out.

The extraterrestrial impact theory, put forward by US scientists Luis Alvarez, Walter Alvarez, Frank Asaro and Helen Michel in 1980, is becoming recognized as the accepted view of the event at the so-called K/T boundary. (K/T stands for Kreide/Tertiary, in which Kreide is the German word for Cretaceous.) The theory suggests that a huge asteroid or meteorite struck the Earth at a velocity of more than 6 miles (10 km) a second. The enormous energy transfer to the Earth's crust and atmosphere probably resulted in catastrophic environmental events such as tidal waves and extensive forest fires.

Some of the more prominent geological evidence to support the theory comes from the so-called iridium anomaly. Iridium is an extremely rare metal in the Earth's crust but in some types of meteorite it is thousands of times more common. The thin rock or clay layer marking the K/T boundary in sedimentary rocks worldwide is rich in iridium, as would be expected if the dust from a pulverized meteorite or asteroid entered the atmosphere and oceans at that time.

The impact on the Earth of an object 6 miles (10 km) in diameter would generate a gigantic explosion – some estimate that the kinetic energy of this explosion would be 10,000 times greater than that resulting from the detonation of the world's nuclear arsenal. The physical consequences of this devastating impact seem to be present in the rocks of the K/T zone. Quartz grains are shattered and marked by severe physical stressing. The vast quantities of tiny glassy spherules – solidified droplets of once molten rock – found in many regions are thought to be the result of extreme temperatures and pressures at the impact zone.

1 Stress on the Earth's crust from a meteorite impact would have triggered massive earthquakes.

2 Tsunamis, or tidal waves, would have been caused by the impact itself and by related underwater earthquakes.

3 Energy released by the impact would have led to widespread forest fires.

4 Particles thrown into the atmosphere would have caused dust fallout over a huge area.

5 Acidic gases produced in the impact would have been spread round the world by air currents, eventually falling as acid rain.

6 Solar radiation would have been blocked by dust and debris, causing rapid cooling – a so-called nuclear winter.

7 An excess of infrared-absorbing gases, such as carbon dioxide, might have led to a "greenhouse" warming effect.

The location of this site has yet to be proved beyond doubt but, of the proposed candidates, the Chicxulub crater off the Yucatán coast on the south side of the Gulf of Mexico seems to have the best credentials (see map). Moreover, the K/T zone around the site – specifically in Haiti and Mexico – is packed with a layer of spherules up to 20 in (50 cm) thick.

Fossil evidence from the K/T zone in the Mexican area reveals tree fragments intimately mixed with material from the sea floor. This is thought to be the result of tidal waves 3,300 ft (1 km) high scouring the coasts of the Gulf of Mexico, mixing land and sea material before dumping it on the sea floor where it was subsequently fossilized.

Studies published in 1992 bolstered the credentials of Chicxulub as the site of the impact. Rocks from the crater's centre have a high iridium content. Argon isotope dating methods suggest that these once molten rocks cooled to solidity some 65.2 million years ago (with a 400,000-year margin of error). In other words, the impact that caused the rocks to melt coincided with the global devastation in the fossil record.

The sheer unpredictability of a huge thunderbolt from space almost certainly swept aside selectively advantageous

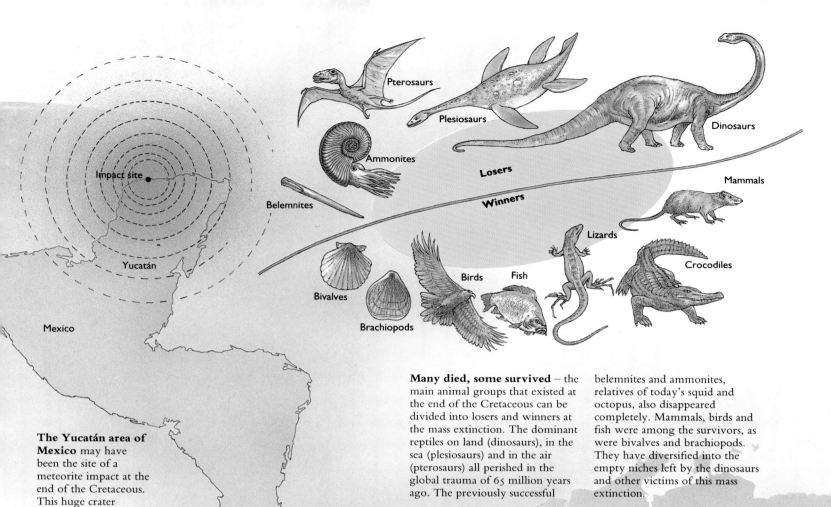

Losers
Pterosaurs
Plesiosaurs
Dinosaurs
Ammonites
Belemnites
Mammals
Winners
Lizards
Crocodiles
Birds
Fish
Bivalves
Brachiopods
Impact site
Yucatán
Mexico

The Yucatán area of Mexico may have been the site of a meteorite impact at the end of the Cretaceous. This huge crater originated at the right time and its 125-mile (200-km) diameter indicates that whatever made it was large enough to have produced effects on a global scale.

Many died, some survived – the main animal groups that existed at the end of the Cretaceous can be divided into losers and winners at the mass extinction. The dominant reptiles on land (dinosaurs), in the sea (plesiosaurs) and in the air (pterosaurs) all perished in the global trauma of 65 million years ago. The previously successful belemnites and ammonites, relatives of today's squid and octopus, also disappeared completely. Mammals, birds and fish were among the survivors, as were bivalves and brachiopods. They have diversified into the empty niches left by the dinosaurs and other victims of this mass extinction.

traits accumulated over the previous hundred million years. Only those groups with a worldwide distribution stood any chance of surviving such a catastrophe – some species belonging to the group must, somewhere, have escaped extinction.

The early mammals, which existed at the same time as the doomed dinosaur species, may have been "pre-adapted" for the rigours of the post-impact world. If their thermoregulation was more efficient than that of the dinosaurs and they could make use of protected underground habitats, they were better equipped to survive the testing K/T times. Thereafter, their rapid diversification into various aquatic, terrestrial and airborne niches must have owed a great deal to the removal of the dinosaurs as the major land vertebrates of the pre-impact world.

A high level of iridium has been found in rock samples from the boundary between the Cretaceous and Tertiary periods. This metal is rare on Earth but much more common in meteorites and its presence gives further weight to the meteorite impact theory. In this picture the thin clay boundary layer is marked by a coin.

JUMPING GENES

Bacteria and blue-green algae are indispensable players on the stage of life. Structurally simple yet biochemically sophisticated, they cause human disease, promote global photosynthesis, fix nitrogen from the air and decompose the bodies of other organisms. But they propagate themselves by asexual cell division, not by sexual reproduction – they do not make gametes nor do they use meiosis for shuffling genes. How, then, do they exploit the raw material of evolutionary change?

Both eubacteria and archaebacteria (pp. 56–57) show staggering degrees of genetic flexibility and variability. This, combined with their rapid multiplication rate (cells divide perhaps every 20 minutes or so), means that they are well equipped to respond quickly via the agency of natural selection when environmental conditions alter.

The genetic flexibility of these organisms derives in part from the basic mutations which occur in all types of DNA (pp. 74–75). In addition to this baseline variation, bacteria have at their disposal a remarkable toolkit of methods

for giving up and accepting genes – or part of genes – from other cells. In these gene movements, DNA physically passes from cell to cell and makes genetic changes in the recipient.

The genes are transferred not only between the cells of a single bacterial species, such as *E. coli*, but also between different bacterial types. This genetic flexibility calls into question the very use of the term "species" for bacteria. If a species is defined as a population of individuals that can breed with each other and not with individuals of other species, then it is hard to encompass bacteria in that definition. Bacteria do not "breed" in any accepted sense and can transfer genes between one "species" and another.

Gene flow is far more unfettered in prokaryotes than it is in eukaryotes. Rather than the straightforward lineages of gene flow in higher organisms, there seems to be a communal pool of genes at the disposal of all prokaryotes, no matter what their type or "species".

The toolkit that individual bacteria can use to tinker with their own genes

operates in two ways. First, the DNA within a single bacterial cell can be shuffled around. Second, genes can be passed from one cell to another.

Internal shuffling comes about when transposable stretches of DNA, known as transposons, play a strange jumping game. A transposon typically consists of several thousand nucleotide bases (in comparison, the total genome of *E. coli* contains about 4 million base pairs) with, at each end, a run of bases called an insertion sequence (IS). The latter acts like dotted lines giving the instruction "cut the DNA here".

Every now and then, a copy of the transposon is made and inserted elsewhere in the bacterial DNA (at a point where the IS sections can fit into base sequences). Because of this "copy-jump-join" movement, transposons have been nicknamed jumping genes. The jump can take the copied stretch of DNA from one point to another on the main circular chromosome of a bacterium, or to one of the additional small DNA circles, or plasmids, found in the cytoplasm of most bacterial cells.

In most bacterial cells there is not only a circular chromosome made of double-stranded DNA, but also one or more DNA circles called plasmids. When the bacterium divides into two daughter cells each carries a copy of the main chromosome and of each of the plasmids. The plasmids act like inheritable mini-chromosomes. In any of these portions of DNA there can be specific regions known as transposons. These have the ability to make copies of themselves and to introduce those copies into other sections of DNA. Transposons can, therefore, "jump" between a plasmid and a chromosome, between plasmids and between different regions of the same stretch of DNA – hence their name of jumping genes.

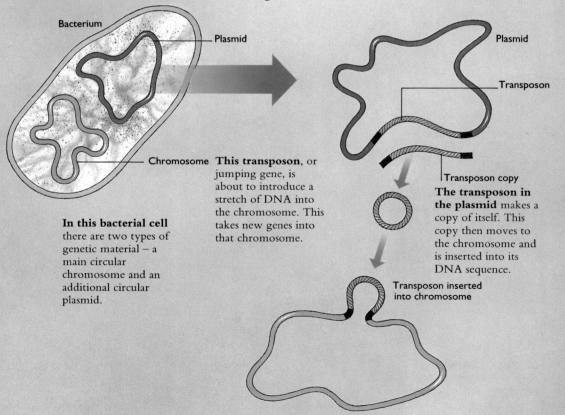

In this bacterial cell there are two types of genetic material – a main circular chromosome and an additional circular plasmid.

This transposon, or jumping gene, is about to introduce a stretch of DNA into the chromosome. This takes new genes into that chromosome.

The transposon in the plasmid makes a copy of itself. This copy then moves to the chromosome and is inserted into its DNA sequence.

Bacterium

Plasmid

Chromosome

Plasmid

Transposon

Transposon copy

Transposon inserted into chromosome

The genes on a bacterium's chromo-some are adequate for usual environ-mental conditions, but the variable repertoire of genes on a plasmid really comes into its own when circumstances change. *E. coli*, for example, lives in feces in the human gut at a constant human body temperature. When the feces reach soil or water in the outside world, the *E. coli* comes into contact with new con-ditions and microbes, many making antibiotics that are lethal to the bac-terium. *E. coli* may be "safe" because one of its plasmids, known as R100, carries genes for resistance to antibiotics such as sulphonamides, streptomycin, chloram-phenicol and tetracycline, as well as to the toxic effects of mercury salts.

Jumping transposons can move genes

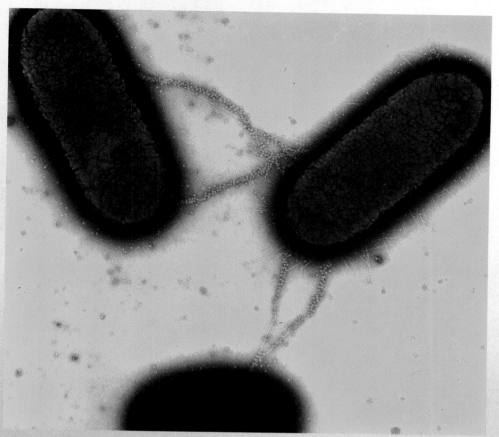

Sex pili join these three cells of the bacterium *E. coli*. The cell at the top right of the picture has the sex pilus gene and has made two pili to each of the other two cells which lack the gene. DNA can pass along these tubes from the first bacterium to the other two. The tiny granules covering the sex pili are a particular type of bacteriophage – a type of virus – that binds to sex pili.

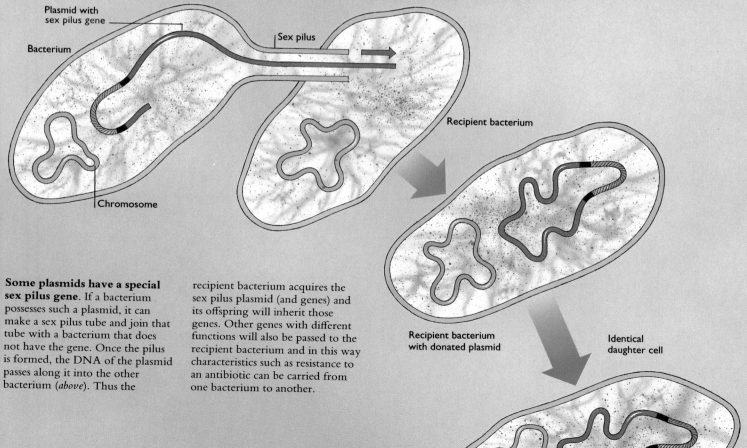

Some plasmids have a special sex pilus gene. If a bacterium possesses such a plasmid, it can make a sex pilus tube and join that tube with a bacterium that does not have the gene. Once the pilus is formed, the DNA of the plasmid passes along it into the other bacterium (*above*). Thus the recipient bacterium acquires the sex pilus plasmid (and genes) and its offspring will inherit those genes. Other genes with different functions will also be passed to the recipient bacterium and in this way characteristics such as resistance to an antibiotic can be carried from one bacterium to another.

from plasmid to plasmid, from plasmid to chromosome or from chromosome to plasmid. In this way both plasmids and chromosomes can swap new gene types produced by mutation.

Of even greater evolutionary significance are a plasmid's fertility genes, which work to promote gene flow between bacterial cells and can sometimes allow transfers from one "species" to another. The fertility genes induce the formation of sex pili – narrow tubes that stick out from a bacterial cell and link with the surface of a cell that lacks the fertility genes. Parts of the first cell's plasmids are then transferred along the pili. Conjugation, as this process is called, is the closest bacteria get to real sex. It has enormous medical importance because the plasmid transfers can pass on resistance to a particular antibiotic from one bacterium to another.

Since the discovery and development of antibiotics, plasmids in disease-causing bacteria have evolved rapidly. Plasmids in bacteria strains isolated before the 1940s, when antibiotics were first used clinically, lacked genes for antibiotic

resistance. By the 1950s, similar plasmids started to carry one or two genes for antibiotic resistance. In the 1990s, it has become usual to find plasmids containing four or five such genes. Bacterial genomes have therefore adapted rapidly to the selection pressure that the products of pharmaceutical companies have imposed on them.

Conjugation, however, is only one of several ways bacterial cells can switch genes. In a gene transfer process known as transduction, bacteria make use of the viruses that attack them. Each of these bacteria eaters, or bacteriophages (a term often shortened to phage), is a bundle of genes surrounded by a protein coat, which can recognize and stick to the cell wall of an appropriate bacterium.

The phage injects its DNA into the hapless bacterial cell and at once forms a host/parasite relationship. The viral DNA induces the bacterial enzymes to make multiple copies of itself and then to make messenger RNA, as a prelude to constructing viral coat proteins on the bacterial ribosomes. Finally, after sets of viral DNA are packaged into the coats,

several hundred new viruses break out of the host cell, killing it in the process. Far from being gene transfer between bacteria, this is, as it appears, bacterial destruction.

But two eventualities can turn this destructive sequence to evolutionary good effect. In the first of these, virus DNA can pick up bacterial genes. This happens when a virulent virus kills the host cell but picks up, at the same time some gene-carrying fragments of chromosomal or plasmid DNA from the doomed bacterium. The resulting hybrid phages have a genome which is partly viral, partly bacterial.

In the second eventuality, a hybrid phage fuses with a bacterial cell which it cannot destroy. In these circumstances the injected viral DNA just inserts itself like a transposon into the DNA of the bacterium. This stretch of new DNA from the virus is called a prophage and will be copied into all the bacterium's progeny. If the initial hybrid virus carried useful genes, such as those for antibiotic resistance for example, this transfer or transduction will change the

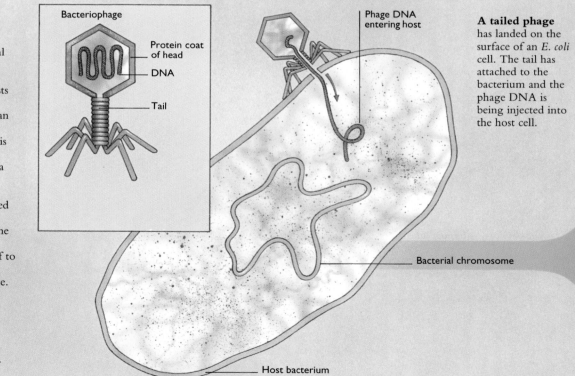

Bacteriophages, often known as phages, are viruses that invade bacterial cells and reproduce inside them. Each virus particle consists of a central gene complement that can be made of either DNA or RNA. This is surrounded by a protein coat called a capsid. The protein coat is particularly complex in the tailed phages that attack *E. coli* cells. With the help of the tail, the phage attaches itself to the bacterium and penetrates its surface. The phage's DNA then passes into the bacterium. Here it may start a lytic sequence or a prophage sequence.

Bacteriophage

Protein coat of head

DNA

Tail

Phage DNA entering host

A tailed phage has landed on the surface of an *E. coli* cell. The tail has attached to the bacterium and the phage DNA is being injected into the host cell.

Bacterial chromosome

Host bacterium

bacterial genome in a beneficial way.

Finally, there is the crudest DNA delivery story of all. At certain stages of their division cycle, bacteria can physically take up lengths of DNA from their environment. Such sequestered DNA can be incorporated into either the chromosome or the plasmids. This direct genetic change, called transformation, can be brought about in the laboratory but is also thought to happen in high-density bacterial colonies where many cells are in the process of dying, breaking down and continually releasing DNA fragments.

With jumping genes, add-on plasmids and gene transfers via sex tubes, virus-powered injection or bulk DNA delivery, it is not surprising that bacteria have gained an accurate reputation as the quick-change artists of evolutionary adaptation. Such capacities, coupled with an extraordinary biochemical alchemy, make the achievements of modern medical science in designing effective drugs to combat the agents of bacterial diseases all the more astonishing.

THE HIV VIRUS

The human immuno-deficiency virus (HIV) infects human cells and ultimately causes the disease AIDS. It is one of an unusual group of viruses, the retroviruses, which have genes made of single-stranded RNA rather than DNA. When HIV enters a human cell – usually a white blood cell called a T helper cell – its RNA is converted into DNA by the enzyme reverse transcriptase and taken into the genome of the human cell. This viral DNA is copied each time the host cell divides.

Disease erupts up to 10 years later when the hidden viral DNA is activated and a new crop of viruses is produced. As the T helper cells are decimated, all the killer infections of AIDS are "let in" by a hopelessly damaged immune system.

An ingenious strategy ensures maximum spread of the virus. Transmission, usually sexual, is direct and efficient since the viruses never have to enter the outside world. The long dormancy period ensures that the infected person unwittingly passes the virus on to other sexual partners. If the virus caused rapid death, its chances of transmission to other hosts would be much reduced.

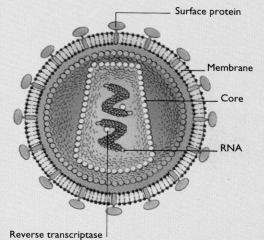

Surface protein

Membrane

Core

RNA

Reverse transcriptase

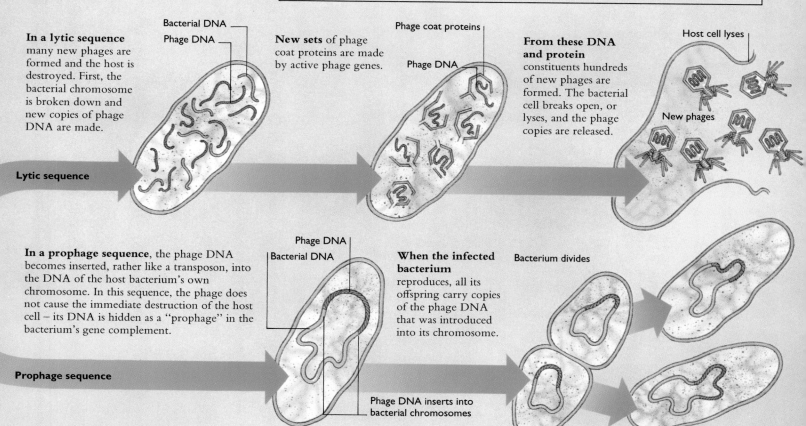

In a lytic sequence many new phages are formed and the host is destroyed. First, the bacterial chromosome is broken down and new copies of phage DNA are made.

Bacterial DNA
Phage DNA

New sets of phage coat proteins are made by active phage genes.

Phage coat proteins

Phage DNA

From these DNA and protein constituents hundreds of new phages are formed. The bacterial cell breaks open, or lyses, and the phage copies are released.

Host cell lyses

New phages

Lytic sequence

In a prophage sequence, the phage DNA becomes inserted, rather like a transposon, into the DNA of the host bacterium's own chromosome. In this sequence, the phage does not cause the immediate destruction of the host cell – its DNA is hidden as a "prophage" in the bacterium's gene complement.

Phage DNA
Bacterial DNA

When the infected bacterium reproduces, all its offspring carry copies of the phage DNA that was introduced into its chromosome.

Bacterium divides

Phage DNA inserts into bacterial chromosomes

Prophage sequence

CONTROLLING DEVELOPMENT

Bacteria may be uninteresting to look at but when it comes to cell metabolism they are ingenious biochemists. On the other hand, comparisons between a seaweed, an oak tree, a worm and an elephant reveal that eukaryotes possess a rather stereotyped metabolism but an astonishing structural diversity.

Most types of bacterial cell look much the same because they are locked inside a simple, single-celled body plan with little variation in their organelles. The individual cells of eukaryotes are not only much more complicated internally but when they are grouped together they can be differentiated into many cell types, opening up limitless structural possibilities. First, eukaryotes have "invented" sex cells. Second, they have evolved the remarkable genes that control the differentiation of cells in the development of embryos.

Multicellular organisms evolved sexual reproduction, which brings about gene shuffling during meiosis (pp. 68–69). Meiotic cell division, the production of gametes, and gamete fusion to form a zygote are all cellular events that originated with the evolution of multiple paired chromosomes in early eukaryotes. The gene shuffling made possible by chiasma formation is only plausible if genes are arranged in this way.

These inventions are the selectional advantages that could have accrued as a result of the transition from the bacteria, which have a single chromosome, to eukaryotes, with their multiple chromosomes. This multiplication of chromosomes may have been the result of increasing cell size and complexity, with an increase in the length of the circular bacteria chromosome. Once this single loop of DNA grew to more than a certain threshold length, accurate copying may have become more difficult. If this were true, mutants that split their DNA into more manageable lengths would have been at an advantage.

In these circumstances a nuclear envelope would have conferred an added advantage because keeping the separate DNA sections confined would have allowed synchronous and carefully organized gene copies to be made before cell

Homeobox genes are crucial developmental genes which almost all many-celled animals appear to possess. The genes determine major patterns in an animal's embryological development. They help to control, for example, the differentiation of the segments of the body and their organization from head to tail. This patterning of the early embryo happens in similar ways in animals as diverse as fruit flies and mammals, both of which have segmented bodies.

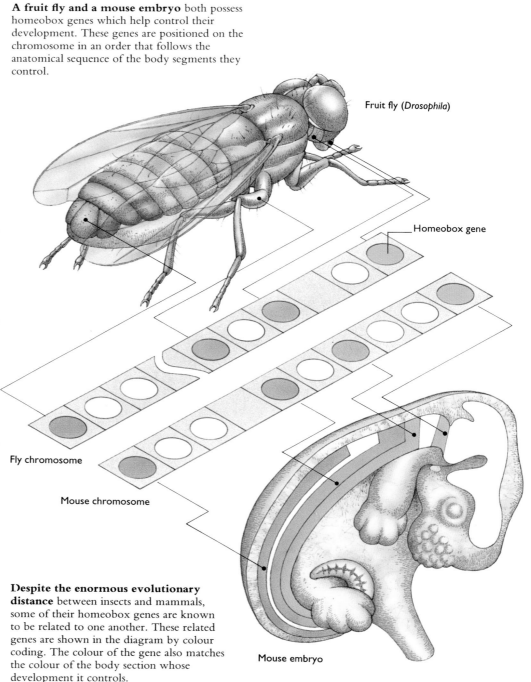

A fruit fly and a mouse embryo both possess homeobox genes which help control their development. These genes are positioned on the chromosome in an order that follows the anatomical sequence of the body segments they control.

Fruit fly (*Drosophila*)

Homeobox gene

Fly chromosome

Mouse chromosome

Despite the enormous evolutionary distance between insects and mammals, some of their homeobox genes are known to be related to one another. These related genes are shown in the diagram by colour coding. The colour of the gene also matches the colour of the body section whose development it controls.

Mouse embryo

division. Whatever the steps that led to the chromosomal scaffolding necessary for sex, the result was clear: evolution was accelerated.

At the same time – probably in the late Precambrian – as sexual strategies were being fashioned and diversity was increasing, genes that controlled body development evolved. Witnessing, today, a fertilized egg develop into an adult creature is like looking into a somewhat distorted mirror of this evolution. The fertilized cell and then its cellular offspring divide by mitosis to form millions of cells. Each of these cells is specialized for its position in the body.

This marvellous unfolding is achieved by developmental genes which organize cell differentiation – the channelling of change which, using different subsets of the same total set of genes, makes one cell become a muscle cell, another a blood cell, another a nerve cell and so on. The developmental genes switch on or switch off appropriate "programmes" of other gene action.

From current knowledge about gene function it seems unlikely that one gene directly influences the activity of another. There are no obvious ways in which one stretch of DNA could interact with another to change its activity. DNA does not work like this – instead it produces its effect by gene products, or proteins. Molecular biologists have now begun to identify developmental genes, the proteins they generate and the ways in which they induce a cascade of differentiation decisions that turns a fertilized egg into a whole creature. The biologists have worked with fruit flies, clawed toads and mice, creatures whose genetic structure is well documented.

The studies reveal a most surprising conservatism in the operation of the developmental genes – the embryonic rules for building a fly are similar to those for constructing a mouse. In fact many of the fundamental pattern-forming processes in these two organisms appear to result from the actions of similar genes. The inference is that this basic kit of developmental genes must have evolved very early for it to be shared by groups as disparate as mammals and insects.

These so-called homeobox genes determine both the shape and the inner construction of a developing organism. Different homeobox genes are active in different regions, thereby promoting the development of an embryo's body. In a sense they are "conducting" genes. One homeobox gene, for instance, ensures that a leg rather than an antenna grows on a thorax segment of a fly. Leg development may require the coordinated efforts of hundreds of genes, but one homeobox gene is enough to initiate it, in the same way as specific movements of a conductor's baton will orchestrate a hundred instruments to play a particular piece of music.

In insects, one of the actions of homeobox genes is to help determine the organs that develop on different body segments. If, however, a mutation stops the action of a specific homeobox gene, such as the gene *BX-C* in fruit flies, this complex organization of segments breaks down. The example below shows what huge effects the mutation of a single developmental gene can exert.

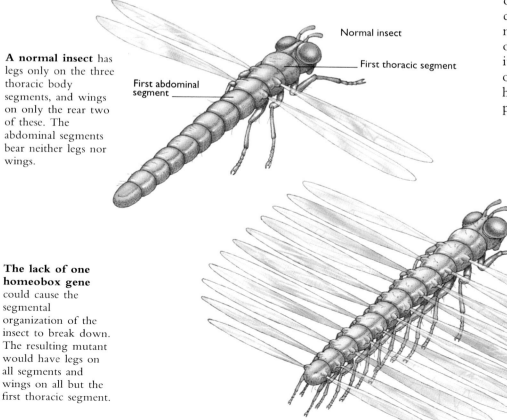

Normal insect

First thoracic segment

First abdominal segment

Mutant insect

A normal insect has legs only on the three thoracic body segments, and wings on only the rear two of these. The abdominal segments bear neither legs nor wings.

The lack of one homeobox gene could cause the segmental organization of the insect to break down. The resulting mutant would have legs on all segments and wings on all but the first thoracic segment.

The proteins generated by homeobox genes are called homeodomain proteins. They do not play a direct part in a cell's metabolism. Instead, their job is to recognize a specific gene sequence by atomic "touch". One end of a homeodomain protein has a helix of amino acids that can sense a specific sequence of nucleotide bases on a gene and then lock on to it. The remainder of the protein can, according to its particular role, block or stimulate the action of adjacent genetic elements.

Via these agents, the homeobox genes can draw marvellous developmental patterns in the raw material of embryonic cells. The examples illustrated on the previous page show how some parts of pattern formation in fruit flies and mice are controlled by homeobox genes.

The nature of this control becomes apparent when a homeobox gene is altered by natural or experimental mutation. A single gene change can have extraordinary and widespread implications. In a fruit fly a change in a single gene makes every segment from the middle of the thorax to the back of the fly try to make a pair of legs and a pair of wings, just as the middle thoracic segment would normally do. Bizarre – and lethal – examples of this sort show what awesome evolutionary potential lies within the development-directing genes. Because of their far-reaching effects a slight mutation in one gene can have dramatic consequences.

But with homeobox genes in place, eukaryotes were effectively shunted to the fast lane of evolution, at least in terms of body form adaptability. The wonderful diversity of today's eukaryotes – from flatworms to ferns, from sea squirts to sequoias – is a measure of the evolutionary vistas opened up by developmental genes.

In the 1980s, the stories emerging from studies on animal homeobox genes stimulated the search for pattern-forming mechanisms in plants. Just as the animal studies relied on a few genetic workhorses such as fruit flies (*Drosophila*) and laboratory mice, plant molecular biologists turned to their green equivalents – the snapdragon, *Antirrhinum*, and

Arabidopsis, a tiny weed belonging to the cabbage family.

In both cases the biologists concentrated on the pattern formation involved in flower production. The "birth" process of a flower starts with multiple dividing cells as opposed to the single fertilized egg that starts a mouse or fruit fly. But once development has begun, the difference is less significant.

A flower develops from a growing point, or meristem, at the tip of a plant shoot where a cluster of cells retain – unlike most other plant cells – the ability to divide continuously. Side clusters bulge outward and form side branches, each with its own meristem.

A tip meristem develops into a flower (not a leafy shoot) when stimulated correctly – often by a change in the relative lengths of day and night. Changing day length signals changing seasons to a plant, enabling it to time flower

production either for optimal growth or for pollination.

On the surface, flowers vary enormously in size, shape and colour, but the underlying plan in the side meristems is normally remarkably conservative: a series of four concentric and consecutively produced rings or whorls.

The first ring, the one nearest the base of the flowering shoot, consists of sepals (bractlike flaps like mini-leaves). Next comes a ring of petals – usually larger than the sepals and typically coloured or shaped, particularly in species that attract insects for pollination. Then comes a ring of stamens, each with a stalk and a double bag of male pollen grains. The central ring lies nearest the shoot tip and consists of carpels which bear the female ovules.

Within each ring, as with each segment of a fruit fly, a particular type of developmental pattern is switched on. In their search for the developmental genes

Analysis of the genes responsible for the arrangement of flower organs suggests that there are three genes switched on in concentric zones. The effects of the gene products on the cells developing in the bud is to produce the correct sequence of organs in the flower. Overlap effects mean that three genes can produce four zones. Two, those of the sepals and carpels, are induced by the actions of single genes. The two intervening zones, those of petals and stamens, appear to be the result of those cells receiving the combined influences of two developmental genes.

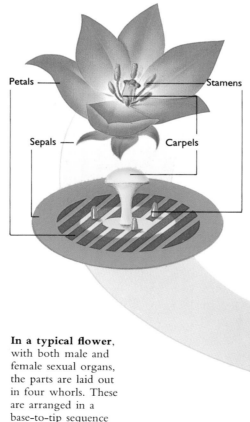

Petals

Stamens

Sepals

Carpels

In a typical flower, with both male and female sexual organs, the parts are laid out in four whorls. These are arranged in a base-to-tip sequence in the flower bud and consist of sepals, petals, stamens and carpels.

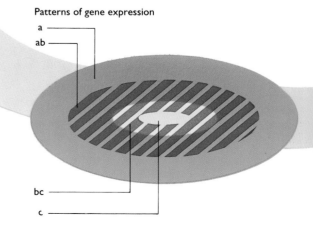

Patterns of gene expression

a

ab

bc

c

that organize this gear switch, the plant gene hunters looked for mutant plants with an inherited defect in flowering ability. By isolating the mutated gene responsible for the defect they could grasp what happens in normal flower production. They discovered that one mutant of this type in snapdragons was due to a jumping gene, which means that transposons (pp. 188–89) are found in eukaryotes as well as prokaryotes. The gene concerned is called *flo*. When it is switched on by external stimuli *flo* initiates the change from a normal shoot tip to a flower. In one mutant that could not respond to daylength stimuli by flowering, a transposon had jumped into the *flo* gene, thereby ruining its capacity to code for a sensible protein.

So *flo* is an important switch gene in the early stages of flowering. At least three more genes, A, B and C, are sequentially switched on during the development of the four flower whorls. If A only is active, then sepals form; A and B together produce petals; B and C together form stamens; and C alone produces carpels. A simple spatial patterning of A, B and C activity will generate the appropriate pattern of flower parts. As with the animal examples, molecular geneticists can dissect out the effects of these control genes by finding strange mutant plants where inappropriate placements and mixes of flower parts are found. The diagrams below show a particular example in which a single gene deletion gives rise to a flower with only sepals and carpels.

The actions of these few control genes which underpin flower development have far-ranging evolutionary implications. Changes in flower structure can prove advantageous for a plant species in different conditions. In one circumstance, for instance, normal hermaphrodite flowers (with both male stamens and female carpels) are advantageous, while in other conditions natural selection might favour plants with a mix of carpel-only and stamen-only flowers.

The developmental genes can, at a stroke, produce massive potential alterations with enormous adaptive significance. Natural selection can, via these genetic agents, shape the simple buttercup or fashion the dazzling complexities of an orchid.

Carpels

Sepals

Gene b inactivated

c

a

Just one de-activated gene in this flower bud means that cells in the second whorl, which should produce petals, get the single gene signal for sepals. The cells in the third whorl, which should produce stamens, receive the single gene command for carpels. The mutant flower that results has only sepals and carpels. No petals or stamens are formed.

When the central gene of the three-gene sequence that controls flower organ production is inactivated by a harmful mutation, the overlap influences on the developing flower bud cease.

EVOLUTION IN OUR HANDS

A new and challenging era has dawned in the twin worlds of genetics and evolution. The fact that scientists can read the human genome as though it were an encyclopedia, and that geneticists can identify the gene and gene product that tells a flower where to put its petals, means that humans can now intervene in the molecular process of evolution.

The same laboratory techniques that allow scientists to investigate the genes of an organism can at a stroke – and in a precise and directed way – be used to change them. It is now in our power to alter the genetic constitution of a single bacterial cell, a yeast, a tobacco plant, a mouse or even a human being in a purposeful way.

This is a sobering thought. Because evolution is in our own hands, as never before, scientists can now, quite literally, lay the raw material of evolutionary change on a laboratory bench and manipulate it. Yet this new situation is one of degree rather than of kind. Human beings have, since the beginnings of agricultural practices about 12,000 years ago, manipulated the genes of other organisms. They have bred more productive strains of wheat, domesticated animals and created all manner of garden flowers.

Genetic engineering provides the most striking example of the new technology at work. The highly specific products of genetically manipulated organisms can now be made in processes not much more complex than the brewing of beer. All these products are proteins and include human insulin, growth hormone, interferon, tumour necrosis factor and blood clotting factors.

These proteins are all substances with immense medical importance. About 50 million people – around one in every 100 of the planet's population – suffer from diabetes. Many have to inject themselves every day with insulin to counter their body's inability to produce enough of this vital hormone. Until the late 1980s the insulin that they injected was derived from pig or horse pancreases and differed from the human form by one or two amino acids. Today, diabetics can use the

Chromosome

Bacterium

Human cell

Cell membrane

Plasmid

Plasmid cut open

The fragment containing the gene coding for insulin is identified and inserted into a bacterial plasmid – a circular "mini-chromosome" removed from a bacterium. The insertion is made by opening up the circle with the same restriction enzyme that was used to chop the human genes. The hybrid or recombinant plasmid containing the insulin gene is put back inside a bacterium.

Restriction enzymes

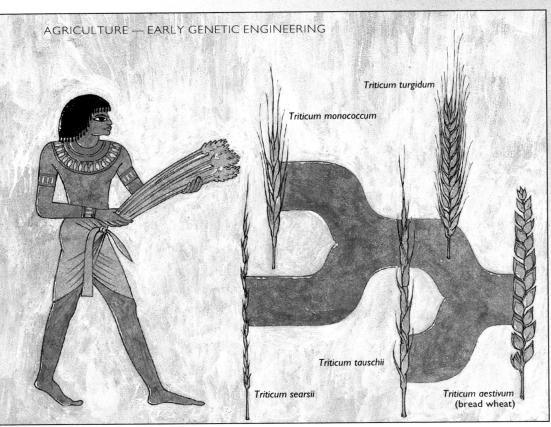

AGRICULTURE — EARLY GENETIC ENGINEERING

Selective breeding – the human-directed analogue of the natural selection that so influenced Darwin's thinking – is a way of purposefully changing the genes of a species. When Neolithic farmers chose, from a range of possible plants in one generation, the seeds to produce the next, their choices were selection in action. Using the clues from the plants' appearance the ancient farmers were able to direct microevolutionary change.

The highly productive bread wheat plant, for example, has been derived from several different species of wild grasses of the genus *Triticum*. These have been crossbred over thousands of years to obtain the most advantageous mix of characteristics. Such agricultural selection is now practised to an ever greater degree by plant breeders using advanced technology.

Triticum turgidum

Triticum monococcum

Triticum tauschii

Triticum searsii

Triticum aestivum (bread wheat)

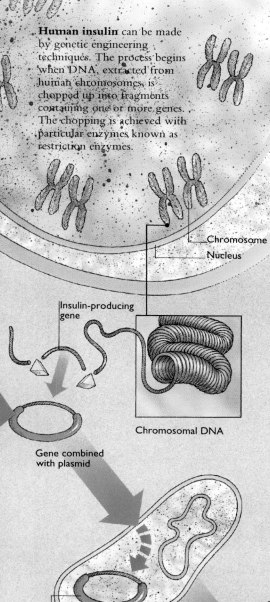

Human insulin can be made by genetic engineering techniques. The process begins when DNA, extracted from human chromosomes, is chopped up into fragments containing one or more genes. The chopping is achieved with particular enzymes known as restriction enzymes.

Chromosome
Nucleus

Insulin-producing gene

Chromosomal DNA

Gene combined with plasmid

Recombinant plasmid reinserted into bacterium

The bacterium with the recombinant plasmid now contains a human gene – the gene for insulin. When placed in an industrial fermentation tank this genetically engineered bacterium will multiply indefinitely, making more cells like itself. All are able to synthesize human insulin. At intervals, the bacteria are harvested from the fermenter and the human insulin is separated and purified for clinical use.

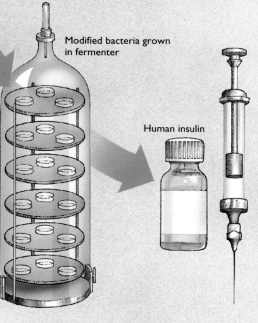

Modified bacteria grown in fermenter

Human insulin

100 percent human insulin that is synthesized in a genetically altered bacterium (see diagrams).

Human growth hormone, previously prohibitively expensive to extract – even in minute quantities – from the pituitary glands of dead donors, can now be manufactured and used to treat congenital diseases that cause stunting in children. Human interferon seems to show great promise in some anti-tumour therapies and in the control of some forms of viral infection.

Genetically engineered clotting factors benefit haemophiliacs because they avoid the risks of viral contamination that may occur when the factors are extracted from blood banks and then transfused into sufferers. Tumour necrosis factor (TNF) can, as its name suggests, cause some forms of cancer to regress. TNF can now be made in fermentation vats in factories.

The marvel of the process is that a bacterial cell, which in more than 3 billion years has never produced a gene to make anything like insulin, can suddenly be persuaded to do just that in a laboratory. Evolution has been short-

circuited because knowledge of DNA and its workings is sufficiently precise to enable a working gene from one organism to be inserted into the DNA of another.

This transfer can take place over an enormous biological divide. Human genes can be fitted into the DNA of a bacterium where, despite the alien environment, they will still work. They will be accurately readable as a code and will still cause a specific human protein to be constructed. It is difficult to imagine a more convincing demonstration of the profound oneness of life on Earth. Despite the eons that have passed since *E. coli* and humans shared a common ancestor, our cells are still able to read each other's genes.

The mechanics of industrial gene transfers are described in the flow diagrams on these pages. Almost every phase of the process testifies to the common ancestry of the genetic machinery of all organisms. The universal nature of the genetic code enables genetic engineers to splice a gene for human insulin into a bacterial plasmid using enzymes (called restriction endonucleases) that bacteria have evolved to chop up the DNA from invading viruses.

Although the procedures used by molecular geneticists to cut out, identify and transfer genes have become commonplace in a laboratory, they place in our hands powers with far-reaching ethical implications. The dilemma lies in where precisely the line should be drawn between activities that are judged as ethically sound and those that can be construed as harmful.

Absolutists would say that all purposeful manipulation of the genes of organisms is unnatural and potentially harmful. This is a difficult position to sustain since selective breeding – the handmaiden of settled agriculture – has manipulated the genes of species for thousands of years. It would be perverse to paint a picture of farming as inherently evil since settled agriculture heralded the start of the modern world.

Some would say it is wrong to tamper directly with the genes of an organism. This activity is different in kind from

selective breeding because it involves the introduction of new genes and raises questions about the release of genetically altered microorganisms into natural habitats. Many would agree that a bacterium with a human insulin gene tucked inside it is a beneficial organism, especially if it can produce a life-saving material more efficiently than by any other route. They might also agree, though, that there should be safeguards to avoid the introduction of such organisms into the wild because, with their changed genetic constitution, they might trigger future microbial evolution along conceivably dangerous paths. In other words, they might end up producing new disease-causing organisms.

The most contentious of all ethical areas concerns the possibility of directly altering the genes of people. The same genetic hardware that enables a human gene to be inserted into a bacterium also provides the technological capacity to introduce genes into human cells. The only context in which this awesome potential has yet been confronted seriously is in the area of "gene therapy".

Many human diseases are the consequence of genetic disorders. Individual people with genes that are either mutated, missing or inactive suffer from a congenital disease that is absolutely inherent in the make-up of their body.

For many of these diseases there are non-genetic ways of intervening to reduce symptoms and enable the sufferer to lead a normal or a near-normal life. In haemophilia A, one of several types of potentially lethal congenital blood clotting disorders, clinicians can replace the missing or inactive gene product that the genetic lesion causes. In this case, the life-sustaining protein called clotting factor VIII can be injected.

For some other congenital diseases, however, this type of intervention is not an option. Cystic fibrosis is a lethal genetic disorder caused by a mutated gene on chromosome 7. This altered gene cannot make a vital protein that enables chloride ions to be correctly transported across the cells lining the lungs. The clinical result is a chronic and progressive failure of lung function

which, despite active physiotherapy and antibiotic treatment, is a cause of disability and death.

The structure of the unmutated gene is known. Is it ethical to try to introduce this gene into the cells of a sufferer's body so that they can make the correct protein rather than the lethal one? Several medical ethical committees around the world have concluded that such attempts are valid and are directed toward the good of the patient.

One such attempt has been to introduce the "good" human gene into a non-harmful virus, which is then inhaled in a nebulizing spray by the patient. The aim of this strategy is to infect the airway cells in the lungs with the virus in the hope that the gene will be taken up along with the harmless viral DNA. The gene could then initiate production of the correct protein so that normal chloride transport could begin. In this case the risks inherent in tampering with a patient's genes must be weighed against the near-certainty of a distressing and terminal illness in the absence of any such endeavour.

A final facet of the ethical imperatives is raised by the fact that viral alteration of the genes of the lung cells would not alter the genes in the sufferer's egg- or sperm-producing cells. Therefore the beneficial genetic changes would not be passed on to subsequent generations. However, such profound alterations are technically feasible.

At a very early stage in a pregnancy a few cells can be taken from the fetus, or from parts of the placenta that are made of fetal cells, and tested for genetic abnormality. In some circumstances, when a congenital disease is discovered and the prognosis is bad for the baby, the genetic evidence is used to vindicate termination of the pregnancy.

Some doctors have argued that if such genetic abnormalities could be detected early enough it would be more ethical to attempt gene therapy to correct the problem rather than kill the embryo. After such therapy, cured patients would pass the introduced genes on to their children. Today, in the early 1990s, we cannot yet come close to carrying out such gene therapy with much likelihood

of success. But the fact that the potential procedure can be described with some precision means the ethical issues posed should be confronted now before the practices arrive.

Darwin and Mendel began, in the last century, an expansion of our insight into the nature of inheritance and evolutionary change. In the middle of the 20th century, Watson and Crick exposed the mechanics of these processes to scientific view. Today, on the verge of the 21st century, those cornerstones of human endeavour have placed in our hands the capacity not only to understand evolution but also explicitly to direct it. Our decisions about how to deploy this ability are just as significant as those relating to our power to destroy the Earth's environment and to use or misuse nuclear energy. Let us hope that we make the critical decisions about directed evolution with humility and wisdom.

A new moral and scientific dilemma faces the world. It is now technically possible to alter, with precision, the genes of any organism – even those of a human being.

The dilemma is symbolized in this composite image (*right*). It shows a baby – a new, genetically unique individual. Surrounding it are chromosomes – the carriers of the genes that made the baby what it is. The glowing yellow spots on some chromosomes are the sites of individual genes illuminated by the use of so-called DNA probes.

Superimposed on the image of the child is the awesome molecule that directed and defined the baby's development – DNA. It is now potentially possible to lay technical hands on that molecule and change it, thus altering the child's genes, the child's body and the inheritance that it can, in turn, pass on to its own children.

With this ability a part of the evolutionary process is within our grasp. It is to be hoped that such devastating power will be used only with the most profound circumspection.

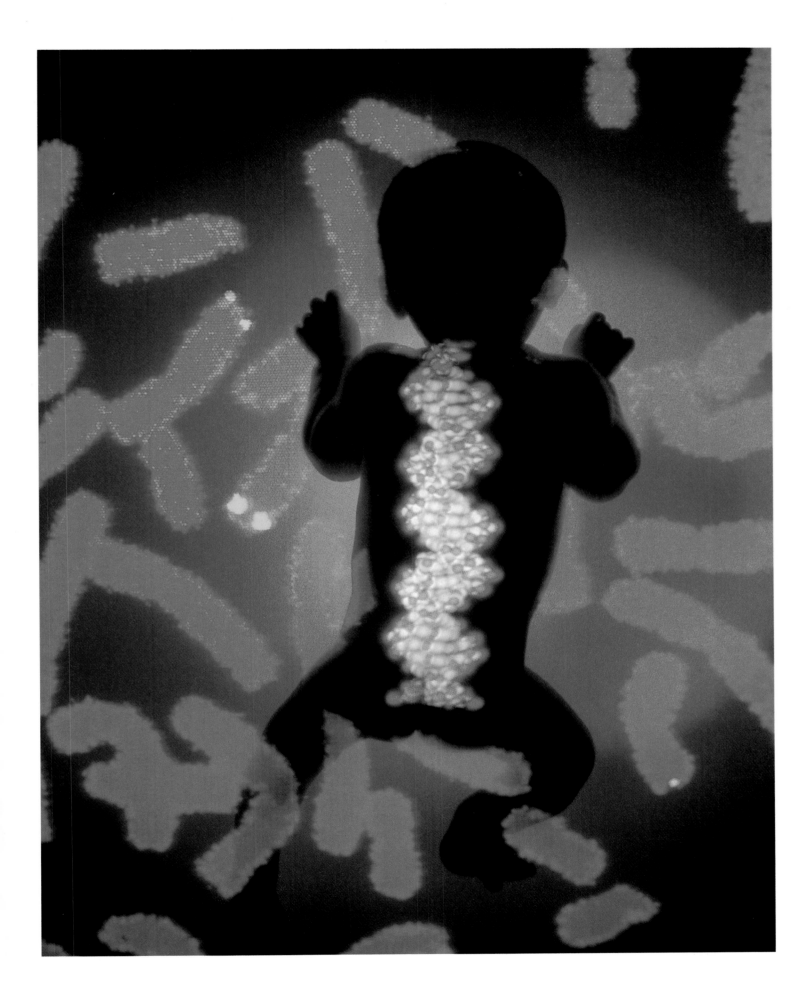

GLOSSARY

Words in **bold** indicate that there is a separate entry under this subject.

adaptation
Modification of the structure, physiology, development or behaviour of an organism which makes it better able to follow its way of life – for example, to live in a particular environment or feed on certain foods. These alterations are achieved by natural selection operating on natural variation within populations of organisms.

By this process, variations within a species are "tested" for their positive or negative impact on the survival and reproductive success of individuals.

adaptive change
A specific and beneficial alteration in the organization, metabolism or behaviour of an organism.

adaptive mutation
A beneficial mutation. Adaptive mutations improve the reproductive success of an organism, either directly by inducing adaptive changes in reproduction or indirectly by altering processes such as food gathering, immune defences or social behaviour patterns.

adaptive radiation
The evolutionary process by which a species or a group of species diversifies into a large number of new species. The new species have more varied lifestyles and occupy a broader range of niches than their ancestors.

For example, the ancestors of mammals have, since the end of the Cretaceous, diversified into a huge variety of animal types, ranging from bats to whales.

aerobic
Describes an organism or a metabolic process, such as respiration, that requires oxygen.

allopatric
Describes the geographical distributions of two species which do not overlap at all. The term, which is opposite to **sympatric**, is mostly applied to species or subspecies that are closely related to one another.

altruism
Behaviour that is advantageous to others. Individuals that appear to forsake immediate advantages (such as the acquisition of food or opportunities for reproduction) so that another individual may benefit are said to display altruistic behaviour.

Such altruism is usually displayed between closely related individuals that share an almost identical genetic constitution. This is usually, in fact, an adaptive strategy: the altruistic individual can ensure that more copies of genes identical to its own pass into the next generation than if it had behaved selfishly.

For example, young white-fronted bee-eaters are more likely to ensure the survival of their genes by helping their parents breed than by reproducing themselves.

amino acid
One of a basic pool of 20 molecules (such as glycine, leucine and methionine) which are joined together in long sequences to make protein molecules. The sequence of amino acids in a particular protein is determined by nucleotide **base** sequences written in the genetic code of **DNA**.

Many other non-typical amino acids, not usually involved in protein production, exist in nature. Examples of these are some of the toxins produced by plants to deter herbivores.

ammonites
Ancestors of squid and octopus that became extinct at the end of the Cretaceous period. Ammonites had chambered, generally coiled, shells.

anaerobic
Describes an organism or a metabolic process, such as the fermentation of yeasts, that does not require oxygen.

anagenesis
The gradual transformation of one species into another by accumulated **adaptive changes**. With **cladogenesis** it is one of the two basic processes whereby a new species can evolve.

archaebacteria
A kingdom of specialized bacteria which typically prosper in habitats with extreme conditions, such as high temperatures or high salinity. All other bacteria belong to the kingdom **Eubacteria.**

arthropod
A member of the **phylum** Arthropoda, which includes insects, crustaceans and spiders.

atom
A basic, essentially indivisible, unit from which matter is composed. Atoms come in a range of types called elements, from the smallest and lightest – hydrogen – to the heaviest naturally occurring element – uranium.

Each atom, of whatever element, is itself made up of a central positively charged **nucleus** – containing protons and neutrons – and a number of negatively charged lighter electrons which exist in a cloud around the nucleus.

In ordinary atoms of an element, positive protons are matched by an equal number of electrons, making the whole atom neutrally charged. By electronic interactions atoms may join together to form molecules. These electronic interactions are called chemical bonds.

ATP

The high-energy organic molecule adenosine triphosphate. ATP is used by the cells of almost all living organisms for the mediation of chemical reactions requiring energy.

During cellular respiration, a high-energy chemical bond links adenosine diphosphate (ADP) with another phosphate group to make ATP. When the energy from this bond is used in a metabolic reaction, ATP is converted back to ADP. In the cells of **aerobic** organisms ATP is produced in organelles called **mitochondria**.

bacteriophage

A specialized type of **virus** which can only reproduce inside bacterial cells. Often shortened to phage.

base

An abbreviation for nucleotide base. Four nucleotide bases – adenine, thymine, guanine and cytosine – are essential molecules in the construction of **DNA**. In the formation of the double-stranded DNA the bases form pairs – adenine on one strand always links up with thymine on the opposite strand, while guanine always joins up with cytosine. The **genetic code** is written in sequences of three base pairs, or triplets, with each triplet coding for a particular amino acid. In the formation of **RNA**, thymine is replaced by uracil.

behaviour

The processes by which living things interact with the world around them – feeding, moving and reproducing are all forms of behaviour.

The term can be used to describe movements or activities of whole organisms or parts of them. The most complex and rapid forms of behaviour are those performed by animals. Plants generally show simpler and slower behaviour patterns.

bivalve

A member of a group of two-shelled shellfish in the **phylum** Mollusca. Examples include clams, scallops and oysters.

body plan

The structural blueprint of a particular group of organisms. The concept enables those studying evolution to chart the modifications of a group's basic construction through adaptive change.

For example, the basic components of the body plan of a **mollusc**, such as radula, shell and muscular foot, have become modified to form creatures so apparently different as squid and snails.

brachiopod

A marine invertebrate animal with two shells. Also known as lampshells, brachiopods superficially resemble **bivalves** but in fact belong to a separate **phylum**, the Brachiopoda. Millions of years ago, brachiopods were successful in terms of species numbers but only a few species still survive today.

brood parasite

A creature, such as a cuckoo, which induces another species to hatch, feed and protect its young.

calcareous

Structures in rocks or living organisms that are composed of, or hardened with, calcium carbonate – or other calcium salts – are said to be calcareous.

Limestone, chalk and marble are all calcareous rocks. The shells and **exoskeletons** of organisms such as **molluscs**, **brachiopods** and **echinoderms** are also calcareous as are the bones and teeth of vertebrate animals.

Cambrian

See **geological time**.

carbon

The essential element in the construction of the organic **molecules** that make up living things. Each carbon **atom** can make chemical linkages with up to four other atoms, including other

carbon atoms. Carbon dioxide, glucose and urea are carbon-based molecules; chains of carbon atoms form the backbone of large molecules such as starch, **DNA** and **proteins.**

carbon dioxide

A molecule present in small quantities in the Earth's atmosphere. It is produced by every organism that respires aerobically and is used as a source of carbon by every plant that photosynthesizes.

Carboniferous

See **geological time**.

cellulose

A compound composed of many glucose molecules made by most plants as a skeletal frame around their cells. It is the main structural material of plants and is the most common organic **molecule** on Earth.

cephalopod

A free-swimming **mollusc** (such as a squid or octopus) belonging to the **class** Cephalopoda. This class also includes the extinct **ammonites**.

characteristic

A specific aspect, or trait, in the organization of an organism. Characteristics under genetic control, such as plumage colour in birds, may be inherited and are subject to natural selection.

chiasma (pl. chiasmata)

The "crossing-over" point between chromosomes. During the production of sex cells, such as sperm and eggs, these crossing-over points are the places where sections of maternal and paternal chromosomes are swapped.

At the chiasmata the chromosomes are cut, ends are swapped and new hybrid chromosomes rejoin. This makes new combinations of pre-existing genes in the offspring produced by sexual reproduction.

chlorophyll

A green, magnesium-containing pigment used by plants to trap the sun's energy in **photosynthesis**.

chloroplast
A chlorophyll-containing **organelle** responsible for **photosynthesis** in many plant cells.

chordate
A member of the **phylum** Chordata, which includes all animals possessing a notochord (a strengthening rod) at some stage of their development. The phylum includes all vertebrates as well as more primitive groups such as tunicates, or sea squirts, and lancelets.

chromosome
A microscopic body made up of **protein** and gene-carrying **DNA**. In the cells of **eukaryotes**, such as animals and plants, chromosomes exist as pairs in the **nucleus** – one chromosome in a pair coming from each parent. But in non-nucleated **prokaryote** cells, such as blue-green algae and bacteria, there is a single circular chromosome.

Each species of eukaryote has a certain number of chromosomes – humans, for example, have 23 pairs. The organization of a chromosome enables its genes to be copied before **mitosis** so that each daughter cell receives a complete set.

cladistics
The analysis of the evolutionary history of groups of organisms based on the order in which species split off from one another. This order can be deduced by determining which of the species' traits are primitive and which are derived from primitive traits.

A key premise in cladistics is that derived traits shared by a number of species indicate a common ancestry while shared primitive traits do not.

cladogenesis
The process by which a single species splits into at least two new species. See also **anagenesis**.

class
A level in the classification of living things. Most **phyla**, which include organisms that share a common body plan, are divided into several classes.

In the phylum Mollusca, for example, there are several classes of **molluscs**, such as Cephalopoda (squids and octopuses), Gastropoda (snails and slugs) and Lamellibranchia (bivalves such as mussels and clams).

classification
The ordering of species into a hierarchy of non-overlapping categories of greater and greater inclusiveness. The **species** is the only "self-defining" natural level in this hierarchy.

Species are grouped into **genera**, genera into **families**, families into **orders**, orders into **classes** and classes into **phyla**. Phyla are grouped into the final, large-scale categories of living things – the kingdoms of life. There are three kingdoms of many-celled organisms – plants, animals and fungi – and two of bacterial life.

clone
A clone of organisms is a group of organisms produced by the asexual reproduction of a single parent. All organisms in the clone are genetically identical.

The offspring that bud off the side of a *Hydra* are clones of the parent animal just as the strawberry plants produced at the ends of runners are clones of the parent plant.

coelenterate
A member of the invertebrate **phylum** Coelenterata, which includes animals such as sea anemones, jellyfish and corals. Coelenterates have a simple body plan with only two main cell layers. They are known to have existed before the **Cambrian**, suggesting that they represent an early group of multicellular animals.

coevolution
The side-by-side evolutionary history of two different species or groups whose lifestyles demand that they interact extensively and predictably with one another. Each species becomes a sig nificant part of the environment of the other and, as a consequence, the two species evolve reciprocal adaptations.

The fig and the fig wasp, for example, have coevolved to such an extent that one cannot exist without the other. The fig must be pollinated by the wasp while the wasp must lay its eggs in the fig's fruit.

Coevolutionary trends are especially common between host and parasite species and between symbiotic partner species. Termites and the microorganisms in their gut that digest cellulose are one such symbiotic partnership.

colony
A group of organisms of a single species that congregate in one place. Species members that gather together for communal breeding form a breeding colony. Where there is a physical link between the different individual organisms, for example the polyps in a coral, the colony forms a "super-organism".

community
The mix of species that typically live together in a specific habitat within an ecosystem. The community will usually contain species representing all the major levels of an ecological system – photosynthetic primary producers (such as plants), herbivores, carnivores and decomposer organisms.

competition
The contest between organisms for resources within a particular environment. Competition can take place between members of the same species or between those of different species. The competition may consist of direct confrontations over a specific resource, such as food, light, water, living space, nesting sites or mates.

The indirect effects of the removal of those resources may also cause competition. For example, if starlings take over nest holes in trees there may be fewer available for other species, thus creating competition.

continental drift
The gradual movement of landmasses resulting from the shift of the tectonic

plates that form the Earth's crust. This has caused dramatic alterations in the shape of the world's landmasses over time and thus influenced the course of evolution.

convergent evolution
The process by which two or more independent lines of evolutionary development bring about superficially similar end-points.

When different groups of organisms are subjected to the same selection pressure they tend to evolve a similar design feature. Streamlining, for instance, gives great energy-efficient benefits to any animal that swims under water. As a result, sharks, dolphins and the now-extinct marine reptiles, ichthyosaurs, although unrelated, all converged on a similar streamlined body form.

Cretaceous
See **geological time**.

developmental genes
Genes that control key aspects of embryological development. In animals, they help to determine the arrangement of organs, either along an organism's head-to-tail axis or, in the case of segmented animals, on different segments. In flowering plants, they organize the concentric pattern of structures – sepals, petals, stamens and carpels – in a flower. Many developmental genes are known as homeobox genes.

Devonian
See **geological time**.

dicotyledon
A flowering plants whose seeds have two seed leaves, or cotyledons. Plants as diverse as daisies, cabbages and oak trees are all dicotyledons.

differentiation
The process by which different types of specialized cell are produced in many-celled organisms. Every body cell is ultimately derived from a single fertilized cell and is therefore equipped with the same set of genes.

During development, however, different subsets of the total gene complement are activated in different cells. This causes them to develop into cell types with specialized functions, such as muscle cells, nerve cells, sperm cells and liver cells.

diploid
The term used to describe the paired **chromosomes** in each body cell of a multicellular animal or plant. The nuclei of these cells contain sets of chromosomes arranged in pairs. One chromosome in each pair is ultimately derived from the male parent, the other from the female parent. A cell or an organism which contains nuclei with this paired arrangement of chromosomes is said to have a diploid genetic constitution.

Each species has a fixed diploid number of chromosomes (signified by 2N). In humans, for example, 2N = 46.

diversity
The range in variation in body form or other traits in a group of related species.

DNA
Deoxyribonucleic acid, known as DNA, is the long molecule used by almost all organisms to store their genetic information.

The DNA molecule is a double helix in which two helical strands are linked by nucleotide **bases**. The complete sequence of bases, organized into triplets, is the **genetic code**. DNA is found in the single-stranded **chromosomes** of bacterial cells, in the chromosomes within the nuclei of animal and plant cells and in some **organelles,** including **mitochondria** and **chloroplasts**.

dominant gene
A gene whose effects on the **phenotype** or physical constitution of an organism dominate over those of a so-called recessive gene when they are present together at the same points on each of a pair of chromosomes.

The gene for brown eyes in humans, for example, is dominant whereas the

gene for blue eyes is recessive. Where the two occur together the brown dominates.

echinoderm
A member of the invertebrate **phylum** Echinodermata. This marine group includes sea cucumbers, sea urchins, sea lilies, starfish and brittle stars.

Ediacaran
Refers to an assemblage of fossil animals from the late **Precambrian** found in the Ediacara Hills in Australia.

Ediacaran animals are enigmatic many-celled animals with soft bodies and no skeletons. They might be early coelenterates or their relatives, or a separate early experiment in multicellular organization that did not lead to any surviving animal group.

enzyme
A protein which facilitates or speeds a particular chemical reaction in an organism. The enzyme trypsin, for example, which is secreted by the gut, digests proteins in food.

Eocene
See **geological time**.

eubacteria
The kingdom of normal bacteria (as opposed to **archaebacteria**). Many eubacteria live as common decomposer organisms in aquatic and terrestrial habitats or are disease-causing agents.

eukaryote
An animal, plant or fungus made up of cells containing a nucleus. Within the nucleus a set of paired chromosomes contains the organism's **genome**. Inside a eukaryotic cell are a number of assorted **organelles** for different tasks such as **mitochondria**, **ribosomes** and, in plants, **chloroplasts**.

exon
A section of the DNA in a gene that contains the code for the manufacture of a specific sequence of **amino acids**. In many genes the particular exons that together code for the **amino acids** of a

whole protein **molecule** are interspersed by non-coding regions called **introns**.

exoskeleton
The external skeleton found in animals in many invertebrate groups. The stiff covering of most crustaceans and insects, for example, is an exoskeleton.

extinction
The disappearance of an entire species. When the death rate of the individuals in a species exceeds the birth rate for a long enough period, the species becomes extinct. Species can also disappear by evolving into a new species or by splitting into new species.

A mass extinction is the loss over a short period of geological time of a large number of different species. The last mass extinction took place at the end of the Cretaceous, 65 million years ago, when all the dinosaurs died out. Such events are thought to be caused by large-scale, sometimes global, changes in environmental conditions.

family
One of the intermediate levels in the **classification** of organisms. Groups of related **genera** are together regarded as a family. For example, lions, tigers and all other species of cat belong to the cat family (Felidae).

fertilization
The fusion of two sex cells to produce the fertilized cell from which a new organism develops. The sex cells are **haploid** – they contain one set of unpaired **chromosomes** – and combine to form a diploid cell with a full complement of chromosomes.

In complex animals fertilization is usually achieved when a mobile sperm from a male fuses with an egg cell from a female. In higher plants fertilization is achieved by the fusion of a pollen tube with an ovule.

fitness
A measure of an individual's overall reproductive success. Genetic **mutations** can either enhance or detract from this success. An organism that can replace itself in the next generation with many offspring has a higher fitness than one that replaces itself with fewer offspring.

fossil
The petrified remains or imprint of an organism preserved in sedimentary rocks. Hard, resistant skeletal parts, such as teeth, scales and bones, are more likely to form fossils than the soft parts of organisms. Hence there are many more fossils of organisms with hard parts, such as vertebrates, than of soft-bodied organisms such as jellyfish.

founder effect
A proposed mechanism to explain why the genetic composition of a new breakaway population differs from that of the original population. If the founders of the new population are few in number, there is a chance that they may possess an unrepresentative subset of the total **gene pool** of the original population. As they reproduce, the pattern of gene frequencies in the new population will differ from that of the parental group.

gamete
A sex cell which normally has a **haploid** (halved) chromosome set. Most animals produce sexually differentiated sex cells, or gametes – sperm in males, egg cells in females – which fuse at fertilization. Some algae and fungi have gametes which look the same but which have genes that are sexually differentiated.

gametophyte
The part of a plant which develops at a particular stage of its life cycle and bears male and female sex organs. These organs produce **haploid** (halved) sex cells or **gametes.**

The gametophyte can be a separate plant, such as the liverwort-like prothallus stage in the life cycle of a fern or may be just a group of tissues in flowering plants.

gene
A functional section of the **DNA** in a cell's **nucleus**. A gene contains the information to specify the structure of a single **protein**.

Genes are inheritable and are passed down from generation to generation.

gene flow
The movement of genes throughout the population of a species via sexual reproduction. A **species** can, in fact, be defined as those organisms between which genes can flow. Any disruption of gene flow between parts of a population will lead to genetic differentiation and, ultimately, the creation of new species.

gene pool
The total range of gene variants found within a population of a species or in an entire species. A full description of the gene pool would itemize all the gene variants and their relative abundances.

gene therapy
The attempt to cure or benefit people who suffer from genetically generated diseases, such as cystic fibrosis, by introducing versions of "correct" genes into their body cells.

gene transfer
The molecular biological technique for transferring genes from one organism to another. Viruses provide one much-used vehicle for gene transfer although it is possible to "shoot" genes into cells ballistically with a minute "gene gun".

genera
See **genus**.

genetic code
The language of inheritance carried in the genes of an organism. The "words" of this language are formed when the four bases found in DNA are arranged into 64 three-base messages, or triplets.

The code of all living things is concerned with protein construction. All but three of the 64 possible "words" code for the 20 amino acids that cells use to manufacture proteins. Up to six different words may code for a single amino acid. The remaining three code words are "stop" signals. They indicate that the end of a structural gene has been reached.

genetic engineering
The technical manipulation of the **DNA** of organisms. In one technique, for example, a specific **gene** is transferred from the human **genome** to a bacterium where it can induce the production of a human protein such as insulin.

genetic shuffling
The bringing together of varied gene combinations through sexual reproduction. There are two phases of this shuffling. First, as male and female sex cells are formed in **meiosis**, genes are shuffled into new combinations. Then two of these **haploid** or halved gene sets in the sex cells come together to make a new **diploid** set at fertilization.

genetics
The study of **genes** and and the part they play in inheritance. Classical genetic studies involve breeding experiments with organisms such as peas and fruit flies to discover which genes determine which characteristics. Modern molecular genetics also examines the molecular structure of the genes themselves.

genome
The total gene set possessed by an individual organism.

genotype
The genetic constitution of an individual organism. The term is generally used in contrast with **phenotype** – the physical organization of an organism's body.

genus (pl. genera)
A grouping of very closely related species. A species' generic name is the first part of its two part Latin name. For example, *Homo sapiens* belongs to the genus *Homo*, along with the now-extinct *Homo erectus*.

geological time
The history of the Earth divided up into eras, periods and epochs. See table:

ERA	PERIOD	EPOCH	Millions of years ago
PRECAMBRIAN			4,600
PALEOZOIC	Cambrian		570
PALEOZOIC	Ordovician		500
PALEOZOIC	Silurian		440
PALEOZOIC	Devonian		410
PALEOZOIC	Carboniferous		365
PALEOZOIC	Permian		290
MESOZOIC	Triassic		245
MESOZOIC	Jurassic		210
MESOZOIC	Cretaceous		140
CENOZOIC	Tertiary	Paleocene	65
CENOZOIC	Tertiary	Eocene	55
CENOZOIC	Tertiary	Oligocene	38
CENOZOIC	Tertiary	Miocene	25
CENOZOIC	Tertiary	Pliocene	5
CENOZOIC	Quaternary	Pleistocene	2
CENOZOIC	Quaternary	Holocene	0.01

Gondwana
The ancient continental landmass formed when Pangaea broke up about 180 million years ago. Gondwana subsequently split to produce the modern continents of South America, Africa, India, Australasia and Antarctica.

gradualist evolution
Evolution by the gradual accumulation of fitness-enhancing mutations. Often contrasted with the **punctuated equilibrium** model of evolution.

gymnosperm
The word gymnosperm means "naked seed" and refers to a group of plants including conifers, cycads, ginkgos and a few other specialized forms such as *Welwitschia*. Gymnosperms were the first plants to develop seeds.

habitat
The particular environment in which an individual organism or species lives. The environment consists of other organisms as well as physical features and certain conditions of climate. The habitat provides a context in which natural selection "tests" particular inheritable variation in a species.

haemoglobin
An iron-containing **molecule** which carries oxygen in the blood of many animals.

haploid
A term for the halved set of **chromosomes** found in sex cells, as well as in the developmental stages of certain plants and in protozoans.

At fertilization the two halved sets reconstitute the species' paired or **diploid** number of chromosomes.

heredity
The transmission of characteristics from one generation to the next.

histones
Proteins found mainly in cell nuclei. They are involved in the organization of the complex folding and supercoiling of the **DNA** in **chromosomes**.

Holocene
See **geological time**.

homeobox genes
See **developmental genes**.

hormone
An organic **molecule** used as a chemical messenger or signal. In animals hormones control sexual maturation, for example. In plants they control such responses as those to the direction of sunlight.

hybrid
The offspring which are produced when individuals from two distinct species are, unusually, able to breed together, such as a horse and a zebra. In many circumstances the resulting hybrid is sterile.

The term also describes the offspring of two individuals which, although they belong to the same species, have significantly different genetic constitutions.

immune system
The **molecules** and cells of an organism that are concerned with defence against attack by invading pathogens or parasites, and against the presence of cancer cells. Bacteria-eating cells such as macrophages and protective molecules such as antibodies are both crucial parts of an immune system.

inbreeding
The process of a continued sequence of matings between the offspring of a small often isolated group of organisms. Inbreeding generally results in an increase in the genetic similarity between individuals in the group since no new genes are introduced.

intron
A portion of a gene which is transcribed into messenger **RNA** (mRNA) but is not used to code for part of an **amino acid** sequence.

Genes are often composed of many coding regions called **exons** divided by non-coding **introns**. The intron-coded sections of mRNA are removed by **enzymes** before the mRNA dictates the construction of a protein on a **ribosome.** Introns are not present in **prokaryote** genes.

invertebrate
An animal without a backbone. The term includes a huge range of creatures such as insects, crustaceans and molluscs.

jumping gene
See **transposon**.

karyotype
The **chromosome** complement of each cell in an organism. The karyotype includes information about the numbers of chromosomes as well as their relative lengths and shapes.

kinship
The close relationship between individuals of the same species. In practical terms kinship is a relative measure of how much of its genetic constitution one organism shares with another.

lichen
An intimate symbiotic association between a fungus and algal cells. Fungal threads (hyphae) make up the main structure of the lichen and between them are individual algal cells. These may be green algae or blue-green algae. Lichens can grow in very harsh habitats such as on bare rock surfaces.

life cycle
The stages in the life of a species which must be passed through in order to ensure the proper progression from a particular phase in one generation to the same phase in the next.

Life cycles may include a range of larval stages, such as caterpillars and tadpoles, as well as alternating phases of sexual and asexual reproduction.

lineage
An organism's line of evolutionary descent.

macroevolution
The evolution of new species and the large-scale patterns of evolution above species level.

marsupial
A member of a group of non-placental mammals that includes kangaroos, wallabies, wombats and oppossums. A marsupial is born at an early stage of development and continues to grow in a pouch at the front of its mother's abdomen before it is capable of independent existence.

meiosis
The particular type of chromosomal division that takes place during the production of sex cells or **haploid gametes**.

In meiotic divisions **diploid** cells are turned into haploid cells and new gene combinations are produced on single chromosomes by **chiasma** formation – the crossing-over between chromosome segments.

membrane
The surfaces of cells and many cell **organelles**, such as **mitochondria** and **chloroplasts**, are constructed of membranes

Each membrane is composed of a double layer of phospholipid molecules (fatty molecules with a charged end) associated with **proteins**. Some of these proteins protrude from the outer and inner surfaces of the membrane while others traverse the membrane's thickness.

A membrane is only about 10 nanometres thick – it would take 100,000 membranes to make a layer about 1 mm thick.

mesozoan
A simple and probably primitive invertebrate animal. Most forms are internal parasites of other organisms and consist of solid cell masses, usually covered with flagella. Mesozoans do not have a gut, nervous system, blood system or muscles.

metabolism
The multitude of organic chemical reactions that take place in the cells and tissues of an organism. Many of these diverse reactions, such as the breakdown of glucose to provide energy, are controlled by proteins called **enzymes.**

microevolution
The small genetic changes that take place in populations within a single species. These changes represent the replacement of particular **genes** by similar genes already present in low numbers in the population.

Alternatively, new gene variants can arise by small mutational changes.

microfossil
A minute fossil of a microscopic organism, such as a bacterium, blue-green alga, protozoan or single-celled green alga.

Miocene
See **geological time**.

mitochondrion
A double-layered **organelle**, with two concentric membranes, found in the cells of animals, plants and fungi. The prime function of mitochondria is the **aerobic** production of **ATP**.

mitosis
The division of a cell and its **nucleus** into two identical daughter cells. During mitosis a cell's **chromosome** set duplicates itself so that the two daughter cells contain exactly the same genes as the parental cell. Mitosis is the basis of most growth processes and of asexual reproduction.

molecule
Two or more **atoms** linked by chemical bonds. A molecule can be as small as two joined hydrogen atoms or as large as a **DNA** molecule containing millions of atoms of carbon, hydrogen, oxygen, nitrogen and phosphorus.

mollusc
A member of the invertebrate **phylum** Mollusca, which includes creatures such as snails, slugs, clams, mussels, squid and octopuses.

monocotyledon
A flowering plant whose seeds have a single seed leaf or cotyledon. Examples include grasses and lilies. See also **dicotyledon**.

monomer
A small organic **molecule** which can join together with other similar molecules to make branched or unbranched chains. The linked monomers make a **polymer**. Examples include monosaccharide sugars such as glucose which make polysaccharide polymers, including starch and glucose, and amino acids which link together to form proteins.

monotreme
A member of a group of egg-laying mammals that includes the duck-billed platypus and the echidnas. Monotremes are thought to represent an early stage of mammalian evolution.

morphology
The anatomical structure of an organism. Morphology can be studied at a number of levels, from gross anatomy to cell patterns, cell types and the ultrastructure revealed by electron microscopy.

mutation
A change in the nucleotide **base** sequence of **DNA**. Mutations can be caused by external influences, such as toxic chemicals or ionizing radiation. They can also result from mistakes made during the copying of DNA.

Most mutations in the functional parts of **genes** will produce mutant, altered versions of the **proteins** for which those genes are coding. These altered proteins lead to adaptations and changes in characteristics that may or may not enhance the **fitness** of an organism.

natural selection
The key evolutionary process whereby some natural variants of a **species** are favoured because of their high levels of reproductive success.

In certain environmental conditions the natural variants of a species exhibit differing levels of reproductive success. Variants whose characteristics make them most efficient in those conditions will be fitter and produce more offspring.

If the characteristics are inheritable the **genes** that determine them will increase in the population. In this way the environment "selects" between the randomly generated genetic variants that are constantly and naturally produced within a population.

niche
The mode of existence of a **species**. A full description of a niche includes details of all the resources utilized by the species, such as habitat and food.

nitrogen
The commonest gas in the Earth's atmosphere and an element present in many important organic **molecules**, such as **DNA, RNA** and **proteins**.

There are two ways nitrogen becomes part of living things: when plants absorb nitrates from the soil and when bacteria and blue-green algae "fix" it from the air and make it part of organic molecules.

nucleotide
See **base**.

nucleus
The membranous, **chromosome**-containing bag found inside all cells of **eukaryotes**.

Oligocene
See **geological time**.

order
A high-level category in the classification of organisms which is made up of groups of closely related **families**. For example, dogs, cats, mustelids, bears, pandas, raccoons, civets and hyenas are all grouped in the order Carnivora. Orders with a presumed common origin are grouped into classes.

organelle
A structure within a cell, some of which have extremely specific functions. **Mitochondria, ribosomes** and **chloroplasts**, for example, are responsible for **ATP** production, protein synthesis and **photosynthesis**, respectively. Others, such as micro-tubules, are part of larger structures such as flagella and centrioles.

organism
A single living creature. All the cells of a multicellular organism – apart from the sex cells – typically have an identical set of **genes**.

The boundaries between organisms break down in colonial creatures. For instance, a colony of coral polyps may be regarded as either a group of joined individual organisms or one super-organism.

Paleozoic
See **geological time**.

Paleocene
See **geological time**.

Pangaea
An ancient continental landmass which formed about 240 million years ago. At this stage of the Earth's plate tectonic history it seems that all the major continental plates were gathered into a single mass. Pangaea later broke up into two huge landmasses – Laurasia and Gondwana.

parallel evolution
Evolution in which similar patterns of development are present in unrelated groups of species. Examples are the **convergent evolution** of specific characteristics due to design constraints or the creation of a similar range of species filling specialized niches in comparable environments.

parasite
An organism of one species that lives at the expense of an organism of another species. The parasite usually lives inside or on the surface of its host species and often obtains its nutrients from the body tissues or gut contents of its host.

Parasites may be microbes, such as bacteria and protozoans, or much larger organisms, such as worms, lampreys, insects and vampire bats. Brood parasites such as cuckoos induce another species to hatch, protect and feed their young.

parthenogenesis
An aberrant type of sexual reproduction known as "virgin birth", in which unfertilized eggs are able to develop into adult organisms. Aphids are among the creatures that reproduce by parthenogenesis.

Although **genes** may mutate in the production of parthenogenetic eggs, the lack of fertilization in parthenogenesis leads to less genetic variety than in normal sexual reproduction. But parthenogenesis is capable of producing more rapid population growth than reproduction involving fertilization.

phenetic
A type of classification of living things based purely on patterns of structural similarities and differences between organisms.

phenotype
The actual physical constitution of an organism. It is partly generated by the action of the organism's **genes** – its **genotype** – during development and also by external factors such as environmental changes and the availability of food.

pheromone
A chemical laid down or emitted by an animal as a signal, usually to other members of its own species. These signals may be connected with food finding or sexual reproduction.

phloem
A vessel in plants which conducts sugars and other products of **photosynthesis** around the organism. See also **xylem**.

photosynthesis
The production of carbohydrates from carbon dioxide and water using sunlight energy trapped by a pigment such as **chlorophyll**. This metabolic process takes place in **eukaryotic** plants, blue-green algae and photosynthetic bacteria.

phylogenetic
A type of classification of living things based on the phylogeny, or evolutionary history, of organisms.

phylum (pl. phyla)
A high-level category in the classification of organisms. A phylum represents organisms with a characteristic body plan organization that is different from those in all other phyla.

The phylum **Chordata**, for example, includes all animals which possess a notochord at some stage in their development and encompasses creatures as diverse as lampreys and humans.

placental
A mammal, such as a human, in which a placenta provides nourishment for developing young inside the womb.

plasmid
A satellite "mini-chromosome" found in bacteria. Plasmids are circular pieces of **DNA** found within bacteria in addition to the main circular **chromosome**.

An individual bacterial cell may have several plasmids. They contain functional **genes** such as those that generate sex pili (connections) between one bacterium and another and those that code for antibiotic resistance. Because they are easily manipulated regions of DNA that can be transferred between bacterial cells, plasmids have a central role in many molecular biological techniques, including **genetic engineering**.

Pleistocene
See **geological time**.

Pliocene
See **geological time**.

pollination
The process whereby a pollen grain reaches the receptive stigma on the female part of a flower. The transfer of pollen between individuals of a particular species is normally achieved by air or water or via the agencies of insects, birds or bats.

polygenetic
A term that describes a characteristic, such as height in humans, which is determined by the combined action of several **genes**.

polymer
A large chainlike **molecule** made by the linkage of many **monomer** molecules.

polymerase chain reaction (PCR)
A technique of molecular biology that enables tiny amounts of **DNA** to be copied over and over again. Using this technique, even gene fragments from fossils can be amplified in sufficient quantities for their base sequences to be analysed.

polymorphism
The phenomenon of multiple, clearly distinguishable varieties of organisms existing within a single species.

Land snails of the genus *Cepaea*, for example, exist in a variety of differently coloured and banded forms.

population
A cluster of individuals of a single species that is to some extent prevented from interbreeding with other such clusters. **Gene flow** between the individuals of a population will be relatively high, whereas gene flow between individuals of different populations will be relatively low.

Precambrian
See **geological time**.

prokaryote
One of the two major types of cell that have evolved on Earth. Prokaryotes were the first to develop and are typified today by the bacteria and the blue-green algae. Prokaryotic cells do not have a **nucleus**. Their main genetic store is a single circular **chromosome**. The cytoplasm of prokaryotic cells is relatively simple with few **organelles** other than **ribosomes**. See also **eukaryotes**.

protein
A long **molecule** made out of a folded sequence of **amino acids**. This sequence is determined by the genetic code of **DNA**. Proteins are the major structural molecules of cells and their components.

protozoan
A single-celled eukaryotic animal. Modern protozoans include amoebae, ciliates and flagellates.

punctuated equilibrium
A particular theory about the pace of evolutionary change. It suggests that much of evolution consists of long periods of evolutionary stasis, during which few (if any) changes take place, interspersed with "punctuations" – periods characterized by dramatic evolutionary changes.

Quaternary
See **geological time**.

receptor
A cell in a sensory organ that responds to a stimulus in the environment. The rods and cones in the retina of the vertebrate eye, for example, are receptors which are sensitive to light.

Additionally, a receptor is a protein located on the outer surface of a cell which responds to the presence of a specific molecule. Receptor proteins on cells, for example, respond to the external presence of hormones.

recessive gene
A **gene** which has little effect on the **phenotyp**e of an organism when it is paired up with a corresponding **dominant** gene. The effects of a recessive gene, such as that for blue eyes, only become apparent in the "double recessive" condition, when both the **chromosomes** of a pair carry the same recessive gene.

replication
The production of a new copy of a **gene** or **DNA molecule.**

reproduction
The production of offspring by an organism. Sexual reproduction generates new genetic arrangements in the offspring but asexual reproduction most often does not.

rhipidistian
A member of a group of extinct bony fish that are related to present-day coelacanths and lung fish. Rhipidistians are probably the ancestral fish that evolved into amphibians.

ribosome
A minute globular structure inside a cell where **proteins** are manufactured. The cells of all prokaryote and eukaryote organisms contain ribosomes, which consist of ribosomal **RNA** and protein.

RNA
Ribonucleic acid, usually referred to as RNA, is a close relative of **DNA.** There are three kinds of RNA – messenger, ribosomal and transfer.

Messenger RNA (mRNA) is a single-stranded nucleic acid that is formed as a working copy of the **genetic code** of a DNA gene. The mRNA moves to a **ribosome** in a cell (itself composed of ribosomal RNA and **proteins)** where it provides the information for making a protein. Individual **amino acids** are provided for this process by transfer RNA (tRNA). These amino acids enable the genetic code messages of the mRNA to be translated into an amino acid sequence in the protein molecule.

sedimentary rock
Rock that has developed by the build-up of accumulated layers of aquatic and terrestrial sediment. This sediment may consist of muds, clays or sands deposited on the seabed, on the bottom of lakes or in swamps. On land the sediments can be wind-blown sands or volcanic ash and dust. Fossils are commonly found in sedimentary rock.

selection
The impact of the environment or human choice on the survival and breeding success of individual organisms. Over many generations, such natural or artificial selection of inheritable characteristics can change the genetic constitution of a population of interbreeding individuals.

selection pressure
A pressure or force that determines the direction of change favoured at certain times and in certain environmental circumstances. Cold climates, for instance, provide a selection pressure in mammals for adaptations, such as extra-thick fur, that conserve heat.

sexual reproduction
The process in which male and female sex cells, **gametes**, fuse to produce the first cell of a new organism. In most species the organisms producing the male and female gametes are themselves physically differentiated into sexes.

Silurian
See **geological time**.

speciation
The process of producing new species.

species
The major hierarchical level in the various systems of **classification**. A species is unique because it is the only naturally and "self-defined" level in such classifications. The individuals of a species breed with one another to produce fertile offspring but cannot, usually, breed with individuals of other species.

Each species is given a two part scientific name in Latin in which the first part defines its genus, the second its unique species title. Thus, *Bellis perennis* (the common daisy) is one species belonging to the genus *Bellis*.

stromatolite
These **Precambrian** fossils are some of the oldest evidence of life on Earth, dating back more than 3 billion years. Stromatolites are rocklike mounds formed by layers of microorganisms, such as blue-green algae, and sediment which compact into dense mats.

structural gene
A gene that codes for either a structural **protein** or an **enzyme**.

symbiosis
An intimate and mutually beneficial association between two or more species. The species have a higher fitness when living together than if they lived apart.

In obligatory symbioses the constituent species cannot live without each other. For example, termites cannot survive without the protozoans and bacteria in their gut which enable them to digest the cellulose in the plant food they eat. The microorganisms can live only in termite guts. In some symbioses known as endosymbioses one species lives inside the cells of another – as green algal cells, for example, live inside the cells of the coelenterate *Hydra*.

sympatric
A term used to describe species that have overlapping geographical distributions. See also **allopatric**.

taxonomy
The study of the **classification** of organisms. One group of organisms, of whatever level in the classification hierarchy, is called a taxon (pl. taxa).

tectonic plate
One of the many large and irregularly shaped sections of the Earth's crust. A plate typically contains a thick region of continental crust which rises above sea level. At the junctions between the plates new crust material erupts as molten rock and solidifies or old crust material is pushed beneath the surface.

The tectonic plates gradually move over the surface of the Earth, causing geological upheaval, mountain building and **continental drift**.

Tertiary
See **geological time**.

therapsid
A mammal–like reptile belonging to the order Therapsida, such as *Dicynodon*. These reptiles probably originated in the early **Permian** and are thought to be the direct ancestors of mammals.

trait
See **characteristic**.

transcription
The conversion of a **base** sequence of **DNA** into an equivalent base sequence of messenger **RNA**.

translation
The conversion of the **base** sequences of messenger **RNA** into an **amino acid** sequence via transfer RNA.

transposon
A section of **DNA** known as a jumping gene. A transposon has the capacity to insert a copy of itself either into another section of the same DNA molecule or into another DNA molecule.

Triassic
See **geological time**.

trilobite
An early aquatic arthropod which resembles a large, flattened wood louse. Trilobites became extinct at the end of the **Permian**.

triplet code
A functional set of three nucleotide **bases** in a **DNA** or an **RNA** molecule. Each triplet codes for an **amino acid** in the manufacture of a **protein.**

variation
The way the **phenotype** of a particular species may vary from one individual to another.

vertebrate
A **chordate** animal with a backbone, such as a fish, amphibian, reptile, bird or mammal.

virus
A simple form of life consisting of a coat of protein molecules surrounding a **DNA** or **RNA genome**. Viruses can only reproduce inside other cells which they parasitize.

xylem
A water-conducting vessel or tube in a plant. Xylem vessels are hollow ducts formed from the cellulose walls of dead cells. They transport water and mineral salts from the roots to the rest of the plant. The accumulation of layers of xylem vessels produces wood. See also **phloem**.

zygote
The **diploid** cell which results from the fusion of a male sex cell, or **gamete,** and a female sex cell at fertilization. As the zygote divides by **mitosis** so a new individual grows.

BIBLIOGRAPHY

Aiello, L. *Discovering the Origins of Mankind* Longman, Harlow, Essex, UK, 1982

Alberts, B. and D. Bray, J. Lewis, M. Raff, K. Roberts, J.D. Watson *Molecular Biology of the Cell* Garland, New York and London, 1989

Arduini, P. and G. Teruzzi *The Macdonald Encyclopedia of Fossils* Macdonald, London, 1986

——*Simon and Schuster's Guide to Fossils* Simon and Schuster, New York, 1987

Attenborough, D. *Life on Earth: A Natural History* Reader's Digest/William Collins/BBC, London and New York, 1980; Little, Brown, New York, 1983; *The Trials of Life: A Natural History of Animal Behavior* Reader's Digest/HarperCollins/BBC, London and New York, 1992; Little, Brown, New York, 1991

Austin, O.L., Jr. *Birds of the World* Country Life Books/Hamlyn, Twickenham, UK, 1987; Western Publishing, New York, 1983

Axelrod, H.R. *African Cichlids of Lakes Malawi and Tanganyika* T.F.H. Publications, Reigate, Surrey, UK, 1973; Neptune City, N.J., 1973

Ayensu, E.S. (ed.) *Jungles* Jonathan Cape, London, 1980; Crown, New York, 1980

Baker, R. *Migration Paths Through Time and Space* Hodder & Stoughton, London, 1982; ——(ed.) *The Mystery of Migration* Macdonald, London, 1980

Berry, R.J. *Inheritance and Natural History* Collins, London, 1990

Bowlby, J. *Charles Darwin: A New Biography* Hutchinson, London, 1990; Norton, New York, 1991

Brown, M.H. *The Search for Eve* Harper & Row, London, 1990; HarperCollins, New York, 1991

Catton, C. *Pandas* Christopher Helm, London, 1990; Facts on File, New York, 1990

Catton, C. and J. Gray *Sex in Nature* Croom Helm, London and Sydney, 1985; Books on Demand, Ann Arbor, Mich., 1985

Cloudsley-Thompson, J. *Animal Migration* Orbis, London, 1978

Cockburn, A. *An Introduction to Evolutionary Ecology* Blackwell Scientific, London, 1991

Cox, Professor C.B. and P.D. Moore *Biogeography: An Ecological and Evolutionary Approach* Blackwell Scientific, London, 5th ed., 1993; Cambridge, Mass., 1993

Cox, Professor C.B. and Professor R.J.G. Savage, Professor B. Gardiner, D. Dixon *Macmillan Encyclopedia of Dinosaurs and Prehistoric Animals* Macmillan, London and New York, 1988

Crick, F. *Life Itself: Its Origin and Nature* Macdonald, London and Sydney, 1981; Simon and Schuster, New York, 1982

Dawkins, R. *The Selfish Gene*, 2nd ed., Oxford University Press, Oxford, 1989 and New York, 1990

Desmond, A. and J. Moore *Darwin* Michael Joseph, London, 1991; Warner Books, New York, 1992

Diamond, A.W. and R.L. Schreiber, D. Attenborough, I. Prestt *Save the Birds* Cambridge University Press, London and New York, 1987

Downer, J. *Supersense: Perception in the Animal World* BBC Books, London, 1988; Seaver Books, New York, 1989

Dorit, R.L. and W.F. Walker Jr., R.D. Barnes *Zoology* Saunders College/Holt, Rinehart & Winston, Orlando, Fla., 1991

Edey, M.A. and D.C. Johanson *Blueprints: Solving the Mystery of Evolution* Oxford University Press, Oxford, 1990; Little, Brown, Boston, Mass., 1989

Ehrlich, P. and A. *Extinction: The Causes and Consequences of the Disappearnce of Species* Random House, New York, 1981

Fincham, A.A. *Basic Marine Biology* British Museum (Natural History)/Cambridge University Press, London and Cambridge, Mass., 1984

Fisher, J. and R.T. Peterson *World of Birds* Crescent Books, New York, 1988

Ford, E.B. *Ecological Genetics* Chapman & Hall, London, 3rd ed., 1971; New York, 4th ed., 1979

Fortey, R. *Fossils: The Key to the Past* Natural History Museum Publications, London, 2nd ed., 1991; Harvard University Press, Cambridge, Mass., 1991

Friday, A. and D.S. Ingram (eds.) *The Cambridge Encyclopedia of Life Sciences* Cambridge University Press, Cambridge and New York, 1985

Fryer, G. and T.D. Iles *The Cichlid Fishes of the Great Lakes of Africa: Their Biology and Evolution* Oliver & Boyd, Edinburgh, 1972

Gamlin, L. and G. Vines (eds.) *The Evolution of Life* Collins, London, 1987; Oxford University Press, New York, 1987

Gibbons, B. *How Flowers Work: A Guide to Plant Biology* Blandford Press, Dorset, UK, 1984; Sterling Pb., New York, 1984

Gould, S.J. *Bully for Brontosaurus: Reflections in Natural History* Hutchinson Radius, London, 1991

——*Eight Little Piggies* Cape, London, 1991; Norton, New York, 1991

——*Hen's Teeth & Horse's Toes: Further Reflections in*

Natural History Norton, London and New York, 1983
——*The Panda's Thumb: More Reflections in Natural History* Norton, London and New York, 1980
——*Wonderful Life: The Burgess Shale and the Nature of History* Hutchinson Radius, London, 1991; Norton, New York, 1990
Gregory, Professor K.J. (ed.) *The Guinness Guide to the Restless Earth* Guinness, Middlesex, UK, 1990
——*The Earth's Natural Forces* Oxford University Press, New York, 1990
Halstead, L.B. *Hunting the Past: Fossils, Rocks, Tracks and Trails: The Search for the Origin of Life* Hamish Hamilton, London, 1982
Hamilton, W.R. and A.R. Woolley *The Hamlyn Guide to Minerals, Rocks and Fossils* Hamlyn, London, 1974;
The Henry Holt Guide to Minerals, Rocks and Fossils Henry Holt & Co., New York, 1989
Heywood, Professor V.H. (ed.) *Flowering Plants of the World* Croom Helm, London, 1985; Prentice Hall, New York, 1985
Hill, J.E. and J.D. Smith *Bats: A Natural History* British Museum (Natural History), London, 1984; University of Texas Press, Austin, Texas, 1992
Huxley, A. *Green Inheritance* Collins/Harvell, London, 1984; FWEW, New York, 1992
James, W.O. *An Introduction to Plant Physiology* Oxford University Press, Oxford, 7th ed., 1973
Johanson, D. and M.A. Edey *Lucy: The Beginnings of Humankind* Granada, London and New York, 1981; Simon and Schuster, New York, 1990
Kimber, G. and M. Feldman *Wild Wheat: An Introduction* The Weizmann Institute of Science, Rehovot, Israel, 1987
Koning, A. *Cichlids and all other Fishes of Lake Malawi* T.F.H. Publications, Portsmouth, UK; Neptune City, N.J., 1990
Lambert, D. *The Cambridge Field Guide to Prehistoric Life* Cambridge University Press, Cambridge, 1985; Books on Demand, Ann Arbor, Mich., 1985
Laithwaite, E. and A. Watson, P.E.S. Whalley *The Dictionary of Butterflies and Moths* Michael Joseph, London, 1975
Leakey, R.E. *Human Origins* Hamish Hamilton, London, 1982
——*Origins: What New Discoveries Reveal about the Emergence of our Species and its Possible Future* Dutton, New York, 1982
——*The Making of Mankind* Michael Joseph, London, 1981
Lewin, R. *Human Evolution: An Illustrated Introduction*

Blackwell Scientific, London, 2nd ed., 1989; Cambridge, Mass., 1989
Lewington, A. *Plants for People* Natural History Museum Publications, London, 1990; Oxford University Press, New York, 1990
Mackean, D.G. *Introduction to Biology* John Murray, London, 1978; International Ideas, Philadelphia, Pa., 1981
Margulis, L. and R. Fester (eds.) *Symbiosis as a Source of Evolutionary Innovation: Speciation and Morphogenesis* MIT Press, Cambridge, Mass., and London, 1991
Margulis, L. and K.V. Schwartz *Five Kingdoms: An Illustrated Guide to the Phyla of Life on Earth* W.H. Freeman, New York, 2nd ed., 1988
Maynard-Smith, J. *Evolutionary Genetics* Oxford University Press, London and New York, 1989
The Mitchell Beazley Family Encyclopedia of Nature Mitchell Beazley, London, 1992
Moore, Professor D.M. (ed.) *Green Planet: The Story of Plant Life on Earth* Cambridge University Press, London and New York, 1982
——*Plant Life* Oxford University Press, New York, 1991
——*The Guinness Guide to Plants of the World* Guinness, Enfield, Middlesex, UK, 1991
Moorehead, A. *Darwin and the Beagle* Penguin, London and New York, 1971; Viking-Penguin, New York, 1979
Norman, D.B. *Dinosaur!* Boxtree, London, 1991; Prentice Hall, New York, 1991
——*The Illustrated Encyclopedia of Dinosaurs* Salamander, London, 1985
O'Toole, C. and A. Raw *Bees of the World* Blandford, London, 1991; Facts on File, New York, 1992
Parker, H.W. *Snakes: A Natural History* British Museum (Natural History), Cornell University Press, London and Ithaca, N.Y., 2nd ed., 1977
Pellant, C. (ed.) *Earthscope* Salamander, London, 1985
Preston-Mafham, R. and K. *Spiders of the World* Blandford, Dorset, UK, 1984; Facts on File, New York, 1984
Raup, D.M. *Extinction: Bad Genes or Bad Luck?* Norton, New York and London, 1991
Ricklefs, R.E. *Ecology*, Nelson, London, 1973; 3rd ed., W.H. Freeman, New York, 1989
Ridley, M. *The Problems of Evolution* Oxford University Press, Oxford and New York, 1985
Sherman, P.W. and J.U.M. Jarvis, R.D. Alexander *The Biology of the Naked Mole-rat* Princeton University Press, New Jersey, 1991
Stack, A. *Carnivorous Plants* Ebury, London, 1979; MIT Press, Cambridge, Mass., 1980

Stanley, M. and G. Andrykovitch *Living: An Introduction to Biology* Addison-Wesley, Reading, Mass., 1982

Stott, P. *Historical Plant Geography* Allen & Unwin, London, 1981; (Paul & Co. Consortium) Concord, Mass., 1981

Strickberger, M.W. *Evolution* Jones and Bartlett, Boston, 1990

Thomas, B. *The Evolution of Plants and Flowers* Peter Lowe, London, 1981

Thompson, K.S. *Living Fossil: The Story of the Coelacanth* Hutchinson Radius, London, 1991

Thurman, H.V. *Introductory Oceanography* Macmillan, New York, 6th ed., 1990

Wallace, A.R. *A Narrative of Travels on the Amazon and Rio Negro* Reeve, London, 1853; Haskell, 1969 (reprint of 1889)

Whitfield, P.W. and Peter D. Moore, B. Cox *The Atlas of the Living World* Weidenfeld and Nicolson, London, 1989; Houghton Mifflin, Boston, 1989

Wills, C. *The Wisdom of the Genes: New Pathways in Evolution* Oxford University Press, London, 1991; Basic Books, New York, 1991

Wilson, E.O. *The Insect Societies* Belknap, Cambridge, Mass., 1971

Wolken, J.J. *Invertebrate Photoreceptors: A Comparative Analysis* Academic, London and New York, 1971

Wye, K.R. *The Illustrated Encyclopaedia of Shells* Headline, London, 1991

Young, J.Z. *The Life of Vertebrates* Clarendon Press, Oxford, 3rd ed., 1981; Oxford University Press, New York, 1991

Zohary, D. *Domestication of Plants in the Old World: The Origin and Spread of Cultivated Plants in West Asia and the Nile Valley* Clarendon Press, Oxford, 1988; Oxford University Press, New York, 1988

INDEX

Entries in the Glossary (pp. 201–11) are arranged alphabetically but not included in the index.
Please refer to the Glossary for additional information.

ACKNOWLEDGMENTS

Photographic credits

t top *b* bottom *l* left *r* right

2/3 G. Brad Lewis/Tony Stone Associates; 6 Bob Fredrick/Oxford Scientific Films; 8 David Hiser/The Image Bank; 9 Vaughan Fleming/Science Photo Library; 10 Collection Down House/Syndication International; 11 Breck P. Kent/Oxford Scientific Films; 12/13 Dr Jeremy Burgess/Science Photo Library; 13 P. Morris/Ardea; 20*l* Sinclair Stammers; 20*r* Vaughan Fleming/Science Photo Library; 21*t* S. Conway Morris/Dept. of Earth Sciences, University of Cambridge; 21*b* Dr B. Booth/G.S.F. Picture Library; 23 Jane Burton/Bruce Coleman; 24 Dr Peter Wellnhofer/Museum für Naturkunde, Berlin; 26/27 G.I. Bernard/Oxford Scientific Films; 27 Claude Nuridsany & Marie Perennou/Science Photo Library; 29 The Natural History Museum; 46 J. Carmichael/The Image Bank; 47 Dr Gopal Murti/Science Photo Library; 48 Lynn M. Stone/The Image Bank; 49 Jon Gardey/Robert Harding Picture Library; 51 Zefa Picture Library; 53 Joan Root/Survival Anglia; 56 Dr Gopal Murti/Science Photo Library; 57 Prof. Luc Montagnier, Institut Pasteur/Science Photo Library; 65 Stephen Dalton/NHPA; 68 Sandra Lousada/Susan Griggs Agency; 68/69 Jacqui Farrow/Bubbles; 70 Mark Mattock/Planet Earth Pictures; 72/73 Jonathan Blair/Susan Griggs Agency; 74 Jackie Lewin, Royal Free Hospital/Science Photo Library; 76/77 Dan Guravich, Photo Researchers Inc/Oxford Scientific Films; 78/79 Matthews/Network; 84 Bill Wood/NHPA; 85 Dr P.W. Whitfield; 90 G.I. Bernard/Oxford Scientific Films; 91 Frank Fournier/Colorific!; 92/93 Alain Compost/Bruce Coleman; 94 Tony Stone Associates; 96 Frank Fournier/Colorific!; 97 John Visser/Bruce Coleman; 98/99 Carol Hughes/Bruce Coleman; 100/101 Jeff Foott/Bruce Coleman; 102/103 Jim Watt/Zefa Picture Library; 104/105 David Hughes/Bruce Coleman; 106 Dr Eckart Pott/Bruce Coleman; 106/107 Peter Johnson/NHPA; 108 Anthony Bannister/NHPA; 110 Jean-Paul Ferrero/Ardea; 111 Robert Hessler/Planet Earth Pictures; 112/113 Cameron Read/Planet Earth Pictures; 113 Judd Cooney/Oxford Scientific Films; 114/115 M.P.L. Fogden/Oxford Scientific Films; 116/117 Tom Ulrich/Oxford Scientific Films; 117 John Downer/Planet Earth Pictures; 118 S. Maslowski/FLPA; 119 Stephen Dalton/Oxford Scientific Films; 120 Michael Fogden/Oxford Scientific Films; 122 Tom McHugh/Zefa Picture Library; 124 Max Gibbs/Oxford Scientific Films; 125 Robert P. Carr/Bruce Coleman; 126/127 Gunter Ziesler/Bruce Coleman; 128/129 Stefan Meyers/Ardea; 130/131 Dr Scott Nielsen/Bruce Coleman; 131 Jen & Des Bartlett/Bruce Coleman; 132 Michael Nichols/Magnum Photos; 133 Michael Holford; 134 Geoff Dore/Tony Stone Associates; 135 Sean Morris/Oxford Scientific Films; 136 Joe van Os/The Image Bank; 136/137 Doug Perrine/Planet Earth Pictures; 138/139 Michael Fogden/Bruce Coleman; 140 Jean-Paul Ferrero/Auscape International; 141 Jen & Des Bartlett/Bruce Coleman; 142 Otorohanga Zoological Society; 144 Michael Fogden/Oxford Scientific Films; 145 I.R. Beames/Ardea; 148/149 Kathie Atkinson/Auscape International; 150/151*t* Clem Haagner/Ardea; 150/151*b* K. Ammann/Planet Earth Pictures; 152 R. Thwaites/NHPA; 153 Stephen Dalton/NHPA; 154 Brian Rogers/Biofotos; 154/155 M.W. Gillam/Auscape International; 156/157 David Curl/Oxford Scientific Films; 157 Peter Steyn/Ardea; 158/159 Michael Fogden/Oxford Scientific Films; 162 Jeff Foott/Survival Anglia; 163 Peter Scoones/Planet Earth Pictures; 165 E.A. Janes/NHPA; 166 John Mason/Ardea; 168 Manfred Kage/Science Photo Library; 169 Peter Menzel/Science Photo Library; 173 Agence Nature/NHPA; 175 Ken Lucas/Planet Earth Pictures; 179 David Simonson/Oxford Scientific Films; 187 Walter Alvarez/Science Photo Library; 189 Dr L. Caro/Science Photo Library; 195 CNRI/Science Photo Library; 199 Peter Menzel/Science Photo Library; 200 Frans Lanting/Zefa Picture Library.

Illustration credits

t top *b* bottom *l* left *r* right

Chapter symbols throughout: Lorraine Harrison
10 Myke Taylor; 12 Myke Taylor; 14 Myke Taylor; 16 Myke Taylor; 17 Bill Donohoe; 18/19 Mainline Design; 22/23 Mainline Design; 24/25*t* Michael Woods; 24/25*b* Mainline Design; 26/27 Mainline Design; 28/29 Bill Donohoe; 30/31 line artwork: Ann Winterbotham, time chart: Gordon Cramp; 32/33 line artwork: Ann Winterbotham, time chart: Gordon Cramp; 33 colour artwork: Steve Kirk; 34/35 line artwork: Ann Winterbotham, time chart: Gordon Cramp, colour artwork: Steve Kirk; 36/37 line artwork: Ann Winterbotham, time chart: Gordon Cramp; 37 colour artwork: Steve Kirk; 38/39 Ann Winterbotham; 40/41 Steve Kirk; 42/43 Ann Winterbotham; 44/45 Frank Kennard; 48 Tony Graham; 49 Robert Gillmor; 50/51 David Wood; 52/53 David Wood; 54/55 David Wood; 56/57 David Ashby; 58/59 Sue Sharples; 60/61 David Ashby; 62/63 David Ashby, background: David Wood; 64/65 David Ashby; 66/67 Bill Donohoe; 68/69 David Ashby; 71 Mainline Design; 72/73 David Ashby; 74/75 Michael Woods, map: Sue Sharples; 76 Michael Woods; 77*l* Michael Woods; 77*r* Robert Gillmor; 78/79 Alan Male; 80/81 Bill Donohoe; 82/83 Bill Donohoe; 84/85 Aziz Khan; 86/87 Bill Donohoe; 88/89 Frank Kennard; 92 David Ashby; 95 Howard Dyke; 96/97 David Ashby; 99 David Ashby; 100 Graham Allen; 101*t* David Ashby; 101*b* Graham Allen; 102/103 David Ashby; 107 David Ashby; 109 Mainline Design; 110 Annabel Milne; 111*t* Annabel Milne; 111*b* Mainline Design; 112 Mainline Design; 114 David Ashby; 117 David Ashby; 118*t* Graham Allen; 118*b* Michael Woods; 121 Howard Dyke; 122 David Ashby; 123*t* Graham Allen; 123*b* Mainline Design; 125*t* Michael Woods; 125*b* Annabel Milne; 128/129 Michael Woods; 131 Michael Woods; 132 Annabel Milne; 135 David Ashby; 136/137 Howard Dyke; 138 David Ashby; 140/141 Michael Woods; 142*t* Ann Winterbotham; 142*b* Michael Woods; 143 Michael Woods; 145 Michael Woods; 146/147 base artwork: Mainline Design, line artwork: Ann Winterbotham; 147 Annabel Milne; 152 David Ashby; 153 Michael Woods; 154 Michael Woods; 155 David Ashby; 157 Annabel Milne; 158 Vanessa Luff; 160/161 David Ashby; 162*t* David Ashby; 162*b* Annabel Milne; 164 Vanessa Luff; 167 *tl* Michael Woods; 167*tr* Colin Newman; 167*b* David Ashby; 170/171 Bill Donohoe; 172 Aziz Khan; 174/175 Aziz Khan; 175 Steve Kirk; 176/177 Aziz Khan; 178–193 Bill Donohoe; 194/195 Aziz Khan; 196/197 Bill Donohoe.

Editor: Jinny Johnson

Art editors: Lynn Bowers
            Patrick Nugent
            Ruth Prentice

Editorial director: Ruth Binney
Text editors: Lindsay McTeague
            Pip Morgan

Researchers: Jon Richards
            Jazz Wilson
Picture research: Zilda Tandy
            Richard Philpott

Production: Barry Baker
            Janice Storr
            Nikki Ingram

Production coordinator: Tim Probart